INTRODUCTION TO NATURAL AND SYNTHETIC RUBBERS

INTRODUCTION TO NATURAL AND SYNTHETIC RUBBERS

D.W. HUKE
B.Sc.

1961
CHEMICAL PUBLISHING CO., Inc.
212 FIFTH AVENUE NEW YORK, N.Y.

Introduction to Natural and Synthetic Rubbers

ISBN: 978-0-8206-0095-6

Chemical Publishing Company:
www.chemical-publishing.com
www.chemicalpublishing.net

First American Edition:

© Chemical Publishing Company, Inc. - New York 1961

Second Impression:

Chemical Publishing Company, Inc. - 2011

Printed in the United States of America

Contents

Plates

Acknowledgements

The author gratefully acknowledges permission to reproduce figures, plates and information from the following sources.

Figs. 2 (and part of the text), 7, 13, and 20 from H. J. Stern, *Rubber: Natural and Synthetic*, Maclaren and Sons, 1954

Figs. 3, 4, data taken from *Rubber Statistical Bulletin*, January 1959

Fig. 5 from *India Rubber Journal*, 1936 **92** 457

Fig. 9 from *The British Journal of Applied Physics*, 1950 **1** 1

Fig. 12 from an original sketch by Dr. G. Egloff, late of Universal Oil Products Co., in *Thorpe's Dictionary of Applied Chemistry*, 4th edition, Longmans, Green and Co., 1937

Fig. 14 from G. S. Whitby, *Synthetic Rubber*, John Wiley and Sons Inc., 1954

Figs. 16, 35 from G. G. Winspear, *The Vanderbilt Rubber Handbook*, R. T. Vanderbilt Co. Inc., 1958

Figs. 17, 22, and 34 from H. Barron, *Modern Rubber Chemistry*, Hutchinson's Scientific and Technical Publications, 1947

Fig. 19; Plates 3 and 4 by courtesy of David Bridge and Co. Ltd.

Fig. 23, reproduced from a lecture—'Basic Instruction in Compounding'—by Dr. B. S. Garvey, which appeared in *Lectures from a course on Basic Rubber Technology*, from lectures compiled by Mr. A. J. DiMaggio in 1955, given by The Philadelphia Rubber Group with Villanova University. Thanks are also due to Dr. Garvey for permission to use part of the material from his lecture

Figs. 21, 29, 30, 31, 37, and 39; Plates 1, 2, and 7 by courtesy of the Dunlop Rubber Company, Ltd.

Figs. 32 and 33 by courtesy of Redland Engineering Ltd.

Fig. 36 from Transactions I. R. I., 1934 **9** 451

Plate 5 by courtesy of The North British Rubber Co. Ltd., and David Bridge and Co. Ltd.

Plate 6 by courtesy of Lastex Yarn and Lactron Thread Ltd.

Table 1, part of the table from *Rubber Journal and International Plastics* 1958, October 18th, p. 584, by courtesy of The Research Association of British Rubber Manufacturers

Finally, to the many people who have helped me I extend my most sincere thanks. I am also grateful to Mr. R. C. Foster for drawing many of the diagrams, and to Dr. R. A. W. Longden for compiling the index.

Introduction

How many people realize the debt that transport owes to rubber? If we think of trains, aeroplanes, buses and coaches, motor cars and even the humble bicycle, all except trains are highly dependent on rubber. The modern car, for example, has more than 500 rubber parts. The tyres, of course, account for the bulk of the rubber involved—about 140 lb.—but there is still from 60 lb. to 120 lb. of fairly small rubber parts, used in electrical insulation, in seals round the doors and windows, in radiator hose, engine mountings, shock absorbers and fender guards, and if the cushions are of sponge rubber this will add considerably to the weight involved. The list could obviously be extended much further.

Perhaps railways are the part of transport where rubber plays its least obvious role. But even here train seats are sometimes of sponge rubber, doors and windows have rubber seals, and rubber flooring is quite common. One or two countries use rubber tyres for train wheels, rubber damping blocks between rails and bridge platforms, and pads between sleepers and rails. The moving parts of couplings and shock absorbers may also be of rubber. It has justly been said that rubber 'cushions' the modern traveller.

This is not the end of rubber's usefulness. In Britain some 54 per cent of all rubber production goes into tyres and tyre products. The next largest single outlet is in footwear, which consumes some 10 per cent of available rubber. Then comes foam or cellular rubber products with about 6 per cent, and wires and cables with 4 per cent. There follow a number of groups using about 1 per cent or 2 per cent each, and these include belting, hose, proofed goods such as rainwear, rubber thread, surgical goods, tiles and flooring, rings, seals and gaskets. Apart from these well-defined groups there are many products which do not fit into groups at all, and these are lumped together and called miscellaneous. This miscellaneous group is, however, the largest user of rubber, apart from tyres, and takes 15 per cent of all the rubber used in this country.

Even after this formidable list of rubber articles it is surprising that there are at least 40,000 different products based on rubber. This amazing versatility of rubber—it can be used with equal ease for elastic bands, pencil erasers and railway buffers—has caused it to be in great demand, and shortages have been felt acutely by everyone. Luckily in this country there have been few serious shortages, but other countries, notably Germany, have suffered both acute shortages and high prices, and in these periods strenuous efforts were made to produce a chemical substitute for rubber.

The extensive research that has gone into finding a substitute for rubber—in which Germany, France, America, Russia and Britain have all played a part—has produced a range of materials which are all like rubber in some degree or other, but which differ among themselves and often quite appreciably from rubber.

It has been agreed to call these materials 'Synthetic Rubbers', and rubber itself is usually referred to as 'Natural Rubber' because it is obtained not from the laboratory or factory, but from a tree. Although the original intention was to find a substitute for natural rubber, very few of these materials can really be called substitutes. Many of them have quite unique properties which enable them to be used where natural rubber cannot be used at all, for example, at high temperatures or where the material must come into contact with oil. Others have somewhat similar properties to natural rubber and can be used instead of it, for example, in making tyres.

The position of synthetic rubbers has only become established, however, since the end of the last war. Before this the synthetic rubber industry was the Cinderella industry. When a shortage of natural rubber occurred it was encouraged, when adequate and cheap supplies of natural rubber were available it was ignored. But since the end of the last war it has become generally accepted that synthetic rubber has a part to play in commerce alongside natural rubber.

Today synthetic rubber plants are springing up all over the world. Britain, after many years without a synthetic rubber industry of her own, has recently re-entered the field with a large plant at Hythe, near Southampton, and several smaller ones elsewhere.

The future is bright for synthetic rubber. More than 3,000,000 tons of rubber were consumed in 1959, of which synthetic rubber made up over a third. But with the expansion of the motor industry the demands for rubber are increasing every year, in all countries of the world. Stepping up the production of natural rubber is a slow business, as it takes seven years before a newly planted rubber tree is able to supply natural rubber regularly. Even if the newly developed high-yielding trees are planted extensively, most authorities are agreed that natural rubber production will not be able to keep up with the requirements.

The solution lies in making greater use of synthetic rubber. In this country synthetic rubber has been imported since about 1944, but the volume has not been very large. In the early days the quality was markedly inferior to that of natural rubber. Today the position has changed. British manufacturers have adequate home-produced supplies of a range of the latest synthetic rubbers available, and their properties and qualities are a great improvement upon earlier varieties.

The object of this book is to explain what these various rubbers are, how they behave, and why they behave as they do. As synthetic rubbers are now being made in this country and will play an important part in the future of the rubber industry, most of this book is devoted to them. How they are made and how they compare with natural rubber is discussed in the appropriate place.

To enable a reasonable comparison to be made between natural rubber and the various synthetic rubbers the subject has been treated from a scientific standpoint, and to keep the size of the book between reasonable limits much technological information has been omitted. Because of the comparison between natural and synthetic rubbers the first section of this book is devoted to natural rubber. In this section the fundamentals of rubber science are introduced, to be developed later in connection with the synthetic rubbers.

This book is intended as an introduction to a complex subject, and as a survey or report for non-technical readers who wish to know something about rubber. The reader who requires further information on a particular point or topic should consult the bibliography at the end of the book.

PART ONE

Natural Rubber

1

The story of natural rubber

The early history

Although the widespread use and development of natural rubber is comparatively modern, rubber was known in very early times. There is some evidence, for example, that rubber was used—notably for play-balls—in Ethiopia, and from there a ball game spread to Egypt. The use of rubber never became widespread in Europe, however, until the nineteenth century, and this may be connected with the distribution of rubber-bearing plants.

Natural rubber is a product of many trees and plants. But it can only be obtained reasonably easily, and in large enough quantities to make its use worth while, from a few types of tree. These trees were not very common in the Old World, but they occurred much more frequently in the New. It is not very surprising, therefore, that rubber has been known and used by the natives of South America for a very long time.

Columbus is usually given credit for having been the first of the modern Europeans to see natural rubber, and it is said that he brought back rubber play-balls which he obtained from the natives of Haiti on his second voyage in 1493–6. This information comes from early Spanish records of about 1535; and most of the early information on rubber comes from Spanish or Portuguese authors describing the wonders of the New World. It is reported that Cortez, the Spanish conqueror of South America, watched a ball game at the court of the Aztec king Montezuma II. According to Captain Fernandez de Oriedo, the official historian of the Indies, writing in

1535, the game was called 'Batey'. It was played between teams of ten or more in a high-walled court with a line across the middle like a tennis net. The object of the game seemed to be to knock the ball—three or four inches in diameter and solid rubber—through stone rings set vertically in the court walls at each end of the line. This scored a goal. There appeared to be a certain amount of religious significance attached to the game, though heavy betting on the result was not unknown.

The historian makes particular reference to the rubber balls, which 'bounce much higher than our balloons'. This is presumably a reference to European balls, which would be of leather and inflated. The Indian ball, if dropped from the hand, bounced up again nearly to the height of the hand. To the early Spanish explorers this must have seemed wonderful, for there was no material known then—and there are very few now—which can approach natural rubber in its ball-bouncing qualities.

A few years later F. J. de Torquemada wrote of a tree which exuded a white liquid when its trunk was struck with a hatchet. The South American Indians collected this liquid in vessels, or, if they had no vessel handy, allowed it to run over them and dry out. On drying out the liquid left a clear brownish film of the tough elastic substance we now call natural rubber. The Indians peeled the pieces off, boiled them in water, and shaped the rubber into balls or other articles. The liquid itself—usually now referred to as latex—could be smeared onto cloth, and the resulting film of rubber made the cloth waterproof.

The Spanish, records de Torquemada, used rubber to wax their cloaks against the rain. 'They were impervious to water, but the sun ruined them.' This is an early reference to one of the grave disadvantages of untreated natural rubber—raw rubber as it is usually called. In the hot sun raw rubber eventually goes sticky and smells badly. Nevertheless, the Spaniards in South America made good use of rubber. They solved the problem of transporting the liquid metal mercury by using hemp or linen bags insulated with rubber, the bags having the additional advantage of being flexible. They impregnated a great many different materials with rubber and made small cloaks, boots, shoes, carriage-tops and outside coverings for many articles.

Very little rubber seemed to filter across to Europe, and most people, including scientists, remained in ignorance of the new material. A start was made in remedying this situation in 1736, when the French Academy of Science sent a survey party to Peru. The object of the survey was to measure an arc of the meridian at the Equator, and the expedition was led by Charles de la Condamine, a noted astronomer. La Condamine was also a keen naturalist, and from Quito in Equador he sent back some samples of rubber to the French Academy. He also wrote a covering note describing how the Indians reported collecting the white syrup of various trees and drying it over fires. Smoking rubber is a process which is still carried out today.

La Condamine also made what must have been an extremely hazardous journey for those days. From Quito on the west coast of South America he crossed the Andes and followed the Amazon down to Para near the mouth of the river on the east coast, a journey of well over 2,000 miles. From Para, la Condamine travelled north along the coast to Cayenne in French Guiana, and there he made the acquaintance of François Fresneau, an engineer employed by King Louis XV of France. Fresneau was also a keen naturalist and interested in rubber. He had seen many samples of natural rubber, brought by Indians from the interior, but he had never seen a tree that produced it. In fact, few Europeans had, for rubber trees were thinly scattered and grew deep in the jungle.

Fresneau spent 14 years looking for the source of natural rubber, patiently searching out and examining many trees before finding the one he was looking for. He carefully noted a description of the tree, and made a pair of boots from its rubber. Then he wrote to la Condamine and the latter read his report to the French Academy. Fresneau's work also gave the first indication that several kinds of trees produced rubber and, later, rubber-bearing plants were found as far afield as Africa and Madagascar. He also reported that freshly dried rubber, or even a freshly cut surface, tended to stick to itself, the joint being quite as strong as the rest of the rubber.

Thus he stumbled across a very important property of rubber, the property of tackiness or 'tack' as it is usually called. In the early days it was more often a disadvantage than an advantage; but later, use was made of it, and today it would be very difficult to build a tyre without it.

Fresneau made another very important contribution to the knowledge of rubber, and thereby assisted in the birth of the rubber industry. At that time rubber usually reached Europe in the form of rubber balls or boots and shoes, and they were exceedingly inconvenient for fashioning into other articles. The milky sap from the tree, latex, was difficult to transport as it quickly fermented if left exposed to the air. A solvent was needed which would dissolve the rubber. The rubber solution could then be applied directly to cloth, or poured into a mould, or even on to a flat surface, so that when the solvent evaporated an easily manipulated film of rubber would be obtained.

There were very few suitable solvents available, and Fresneau discovered that turpentine was the best. The result was not completely satisfactory, as the film of rubber it left after evaporation tended to become very sticky when hot, and hard and brittle when cold. But it was better than nothing, and Fresneau duly sent off a report to his senior minister in the Government service. Within two months, however, two French chemists had reported similar findings.

The beginnings of the rubber industry

With the discovery of a solvent for rubber the way was open for rubber manufacturing to begin. There were still many difficulties to overcome, probably not the least being the popularization of a rather unprepossessing-looking material, and it needed men of foresight and technical skill to develop it. The adventurous French balloonists began to use rubber to cover their balloon fabrics, and in 1803 a small factory was erected near Paris to manufacture elastic bands.

Interest was awakening in the new substance on this side of the Channel. Up to this time rubber was known by a variety of names of which 'gum elastic' was probably the most common. In 1770 Joseph Priestley, the discoverer of oxygen, remarked that gum elastic would rub out pencil-marks from paper, and so the name 'rubber' was born. Probably the first impetus to the rubber industry in Britain came with the beginning of gas lighting in this country. In 1815 there were 4,000 gas lamps in London, and gas undertakings were springing up in each large town. In 1819 Charles Mackintosh, a dye manufacturer of Glasgow, decided to buy the coal-tar by-products of the Glasgow gas works. He was able to sell the pitch, and to obtain this he had to distil off the lighter oil. This lighter oil, called coal-tar naphtha, proved to be an excellent solvent for rubber. Not only that, it gave a much less sticky rubber on evaporation than did turpentine. The film of rubber after evaporation was still a little sticky, but Mackintosh got over that by laying down a film of rubber on some woollen cloth and covering the sticky surface with another layer of cloth.

This was the original Mackintosh waterproof fabric. The result was by no means perfect. The rubber still smelled rather badly in hot weather, tended to become hard and brittle when cold and never lasted very long; but it was a great improvement on anything that had gone before. Mackintosh took out a patent in 1823 and established a factory in Manchester.

One of the first people to apply for a licence to use Mackintosh's discovery was Thomas Hancock. Hancock and Mackintosh probably started their careers in rubber at about the same time. Hancock was a coach-builder in London when he became interested in rubber, and in 1820 he opened a small shop for making elastic tops for gloves, waist-bands and pockets ('to prevent their being picked'). He cut the rubber into thin strips for sewing into the material and thereby collected a lot of waste pieces. Trying to find a way of utilizing them, he too found that newly cut rubber was tacky, and he was able to stick small pieces together. This suggested a solution to his problem. If he could tear the rubber up small enough he might possibly make a flat sheet. There was an obvious need here for a machine, and Hancock promptly designed one.

Hancock's machine, which is the forerunner of the modern

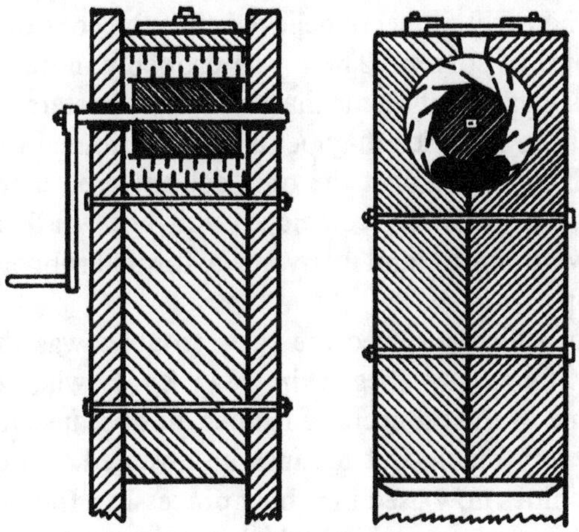

FIG. 1. Hancock's 'Pickle'

internal mixer, was called the 'pickle' to keep the idea secret from competitors (Fig. 1). The machine consisted of a hollow drum lined with teeth. Inside this was a cylinder, also equipped with teeth, which could be revolved by means of a handle. Hancock's idea was to put in a charge of rubber and tear it into small pieces by means of the revolving teeth. He found, however, that it required a great deal more power to turn the handle than he had expected, and the rubber got very hot in the process. When he took it out he found that instead of being in pieces the rubber had turned into a soft plastic ball.

This process is now known as mastication. When the rubber is in this soft and dough-like condition solid powders and even liquids may be mixed into it quite easily. Hancock himself mixed pitch, tar and various other materials into rubber to improve its qualities. This discovery of Hancock's is the basis of the modern rubber industry, for not only does mastication allow various substances to be mixed into rubber, but in the plastic state rubber can be moulded into any desired shape.

Goodyear and vulcanization

Although Hancock and his contemporaries were making great strides in the manufacture of rubber articles, they were not blind to the disadvantages of their raw material. It was very adversely affected by changes of temperature, and in continuous use quickly lost its strength and elasticity.

Many of the early pioneers believed that the root of the trouble lay in the choice of solvent. There was probably more than a grain of truth in this idea, for it seems that the early disasters of the young rubber industry in America may be directly attributed to the use of turpentine, which was almost the only solvent used there. The first rubber boom unfortunately coincided with hot weather; many of the products were ruined, and many newly formed rubber companies went bankrupt.

A company which survived the early troubles was the Roxbury India Rubber Company, begun in 1833 by Edwin M. Chaffee. Chaffee was one of the pioneers of rubber, not because he improved the substance very much, but because he invented two of the most important machines now used in rubber processing. Independently of Hancock he invented a machine which would masticate rubber. It was simply two steam-heated rollers, revolving at different speeds.

Rubber inserted between the rollers was squeezed and torn at the same time and was soon reduced to a thin plastic sheet on one of the rollers. Chaffee found that masticated rubber dissolved in solvents much more easily than rubber which had not been masticated, and he was able to keep his turpentine down to the minimum.

This machine of Chaffee's was the forerunner of the modern mixing mill, and it had the rollers placed horizontally. Chaffee also tried putting the rollers on top of each other, and found that he could coat fabric with rubber in this way. This machine was the forerunner of the modern coating machine or Calender. Chaffee's machine actually had four rollers and weighed nearly 30 tons. It was referred to as the 'Mammoth'.

Chaffee's inventive genius enabled his company to survive many of the early disasters which overtook his competitors, but the company did not prosper. In 1836 the Roxbury Company was approached by an inventor called Charles Goodyear, who suggested a new type of valve for an air cushion. Goodyear had been interested in rubber for some years, but it seems that this was his first contact with the grim realities of the rubber situation. He became interested in the problems of rubber and set out to try and solve them.

Goodyear and Hancock must be considered the two outstanding figures in the history of natural rubber, yet the contrast between them is astonishing. Hancock was industrious and phlegmatic. As each problem presented itself he set to work to solve it carefully and systematically. When he had made his discoveries he patented them, taking great care that they were not easily infringed. He was a good business man and steadily built up his own business, collaborating with Mackintosh when necessary and eventually going into partnership with him.

Goodyear was a visionary. Originally he wanted to enter the Ministry, but was dissuaded from this course by his father who sent him into business in Philadelphia. But Goodyear was never a success in business, and throughout his life trailed a succession of debts behind him. At times he conducted his experiments in prison. Yet in spite of all this and constant ill-health, he retained a conviction that rubber could be 'cured' of the effects of temperature, stickiness and its other adverse properties.

Goodyear realized that improved mechanical treatment of rubber would not solve the problem, and a better choice of solvent was only a partial answer. It seemed that some sort of chemical treatment was the only solution. Goodyear tried mixing into rubber, therefore, everything that he could possibly think of. After many trials, magnesia seemed to give good results, but a spell of hot weather quickly disposed of this theory. Later, boiling rubber with lime seemed to have solved his problem, but once again he was quickly disappointed. At one stage in his experiments he used nitric acid to remove some bronze powder from rubber, and discovered that the acid left the rubber surface hard, leathery and definitely not sticky. Goodyear was convinced that this was the treatment he had been looking for, and he took out a patent for his 'acid gas' process.

Unfortunately the acid gas process did not stand the test of time, because the treatment only took place at the surface of the material and the acid did not penetrate. Nevertheless Goodyear managed to sell some licences for his patent. He sold one to the Roxbury Company and thereby came into contact with a certain Nathaniel Hayward, who was working with the company and who had a patent for mixing sulphur with rubber and exposing it to the sun. Goodyear became interested in this idea, bought the patent, and with Hayward began experimenting and manufacturing. But the effects of the acid gas process and the sulphur treatment proved to be only temporary, and soon they were faced with failure. Unfortunately Goodyear had undertaken the production of a large quantity of mailbags for the United States Government, and the failure of his processes led him into debt again.

After his despair came success. One of his experimental samples, in which sulphur had been mixed with rubber, was accidentally left on a hot stove all night. The rubber charred, instead of melting as ordinary rubber would have done, and Goodyear noticed that it had certain leathery properties. Heat, he reasoned, might be the answer. If he could heat his rubber-sulphur mixture for just the right length of time it might give the change he was looking for. After more experiments Goodyear found that it did—and he was surprised by the extent of the change. He had begun with raw rubber which, though it had a certain amount of elasticity, was generally poor in physical properties. His sulphur and heat treatment converted it into a much stronger material, with much more elasticity, and, more important as far as he was concerned, it remained largely unaffected by changes of temperature and did not become sticky.

Having discovered the secret of changing or 'curing' rubber, he could make no one believe him, and for some reason best known to himself Goodyear did not patent the discovery until five years after he had made it. His patent was issued in 1844.

In about 1842 Hancock received some pieces of rubber which had been treated by an unknown process. The rubber was impervious to heat, did not become sticky and was not much affected by solvents. The pieces were brought from America by the agent of the inventor, who was, it must be assumed, Goodyear.

Hancock was interested in the rubber and he began experimenting. He says that the pieces he examined smelled slightly of sulphur and presumably he used sulphur in his experiments. He also found, as Goodyear had done, that the last decisive step—that of using sufficient heat—was the most difficult to discover. But after many experiments he did discover it, and by immersing rubber in molten sulphur he was able to reproduce the change effected by Goodyear. He also noticed that some of his pieces of rubber, which were left

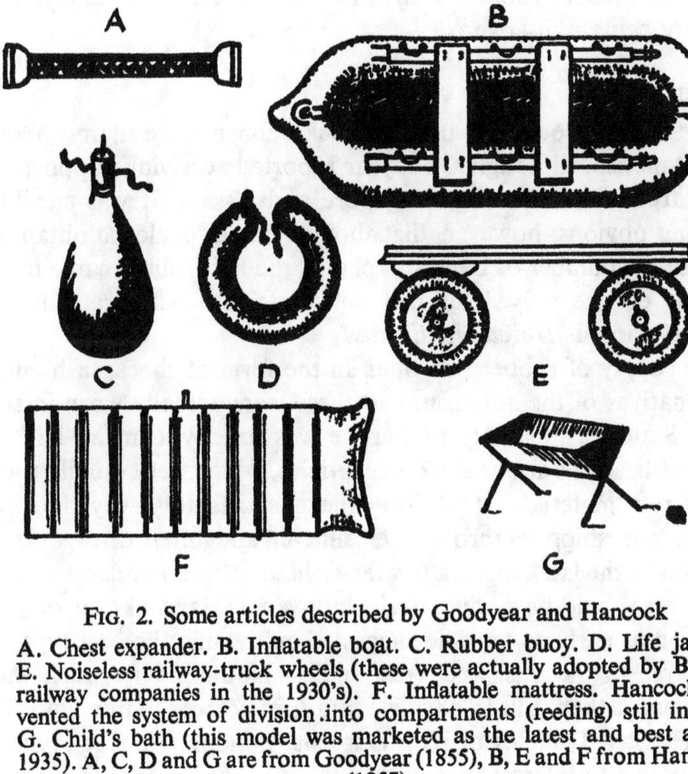

FIG. 2. Some articles described by Goodyear and Hancock

A. Chest expander. B. Inflatable boat. C. Rubber buoy. D. Life jacket. E. Noiseless railway-truck wheels (these were actually adopted by British railway companies in the 1930's). F. Inflatable mattress. Hancock invented the system of division into compartments (reeding) still in use. G. Child's bath (this model was marketed as the latest and best about 1935). A, C, D and G are from Goodyear (1855), B, E and F from Hancock (1857)

longer in the molten sulphur and nearer the bottom of the pot—that is nearer the fire—became hard and brittle. Thus he discovered what is now known as Ebonite. Hancock's patent was issued in 1843. The process was called 'vulcanization' by one of Hancock's colleagues, and the name was adopted generally.

It is clear that Goodyear first discovered vulcanization, but Hancock secured the English patent rights to it, and so Goodyear lost financially and in recognition. Goodyear's story after this is but a continuation of what went before. He patented over 200 applications of rubber and died in 1860, in New York, leaving his widow debts of over 200,000 dollars. The great rubber company which bears his name, the Goodyear Tire and Rubber Company, was formed 38 years after he died.

Hancock, on the other hand, died in 1865 a comparatively wealthy man. He, too, had a long list of patents and inventions to his credit. Between them these two men discovered an astonishing number of uses for rubber. Many of them are still in use today, and many are being rediscovered.

Plantation rubber

In the time of Goodyear and Hancock rubber came mainly from South America, although Goodyear reported receiving supplies of rubber from Java, Penang, Singapore and Assam. It was steadily becoming obvious, however, that though rubber could be obtained from quite a number of trees and plants, the best rubber came from one type of tree only. This tree grew in South America, and its botanical name is *Hevea braziliensis*.

The supply of rubber was often in the form of shoes, fashioned by the natives of the Amazon, and these were actually worn in the United States. The supply to Europe was usually a mixed bag of shoes, rubber balls and many other articles, which were usually used only as raw material. Apart from the unsatisfactory way the raw material was shipped abroad, the difficult and often disorganized collection of the latex made a few far-sighted people wonder whether rubber trees could be systematically cultivated. Hancock was one of these, and as early as 1835 he suggested cultivating the best type of rubber tree in the East and West Indies. Nothing very much was done about it then, but in 1855 we find him corresponding with Sir James Hooker, Director of Kew Gardens, London. Hooker eventually managed to interest an assistant secretary at the India Office in the idea, and H. A. Wickham, who had something of a reputation as an explorer, was commissioned to obtain seeds of the rubber tree from the Amazon valley.

H. A. Wickham, or Sir Henry Wickham as he later became, was already an experienced rubber trader and had sent drawings of the rubber tree (*Hevea braziliensis*) leaf and fruit to Hooker. He also knew that the big oily seeds of the Hevea tree quickly lose their vitality, and he suggested sending young plants. But space and expense were against this and he eventually agreed to supply seeds at £12 10*s*. a thousand.

Soon afterwards, at Santarem, his base near the Amazon river, Wickham learned that the steamer *Amazonas*, on her first trip to the Amazon from Liverpool, was lying downstream with no return cargo aboard. He had no money, but on the strength of the order from the Indian Government in London he wrote to the Captain chartering the vessel and fixing a meeting-place where the Tapajos river meets the Amazon.

He then set off into the Brazilian jungle, and with the aid of natives began collecting the seeds. After collection they dried the seeds, wrapped them in banana leaves and packed them in wicker-type crates which the village women made from local cane. The crates were piled into canoes, and then began the race against time. When they met the steamer Wickham and his helpers unloaded the canoes as quickly as they could, and the steamer made all speed downstream towards the coast.

They had to stop for clearance papers at Para, and what went on there has been the subject of much controversy. It has been said that since the Brazilian authorities would not have allowed a boat-load of their precious rubber seeds out of the country without a certain amount of argument, Wickham deliberately deceived them about the nature of the cargo. On the other hand this has been denied by those who say that Wickham acted correctly throughout. Whatever the circumstances, however, it must have been an anxious time for all concerned, with the perishable seeds aboard, until they were finally cleared. But cleared they were, and made a speedy and uneventful trip to Liverpool.

The *Amazonas* docked at Liverpool with 70,000 Hevea seeds, and immediate preparations were made to rush them to Kew and plant them in hot-houses. Only 2,397 seeds germinated, but a year later (1876) nearly 2,000 young plants were shipped to Ceylon, and later some were sent to Singapore and Buitenzorg. Out of these few seeds a vast plantation industry grew up. But compared with the dispatch with which the seeds had been collected and grown, further progress was exceedingly slow. Experimental and administrative delays combined to hold back production, and it was 25 years before even a small amount of rubber was exported. In 1899 the amount of rubber

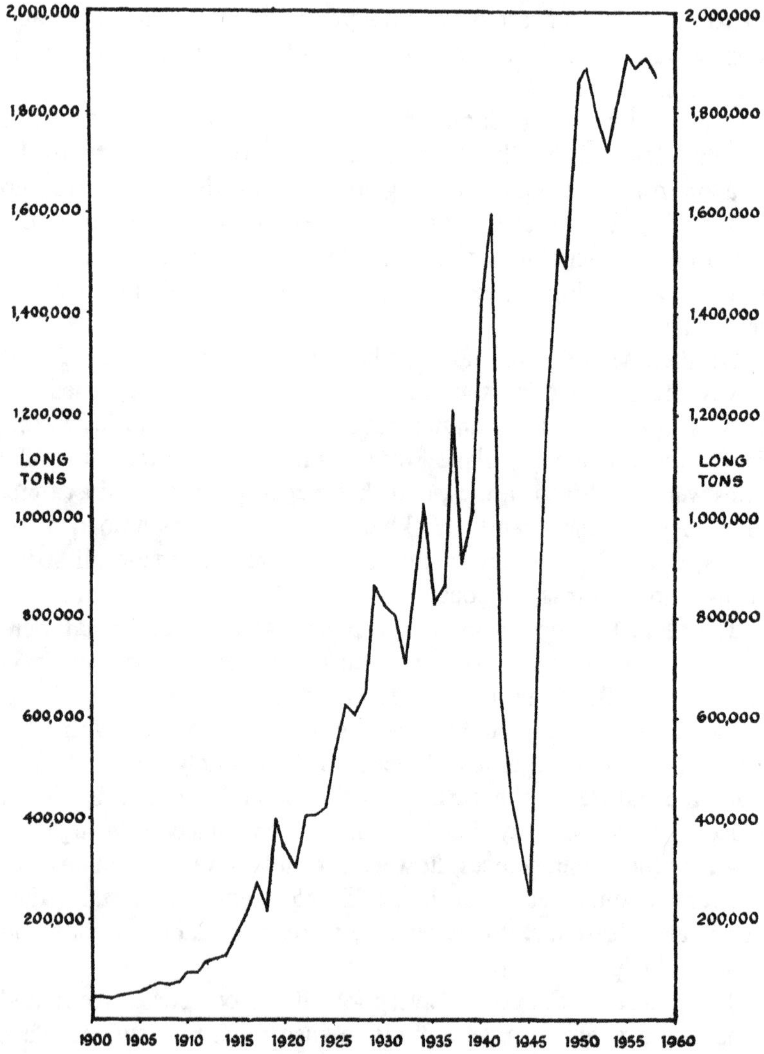

FIG. 3. World production of natural rubber since 1900

exported from Ceylon was only just over three tons. In 1905 the total production from all the Far East plantations was 145 tons. But help for the young plantation industry was near at hand.

The pneumatic tyre had been invented in 1845 by R. W. Thomson, but since the horse-drawn carriages of those days travelled comparatively slowly, the advantage of an air-cushion wheel to reduce friction between the wheel and the road was not fully appreciated, and the

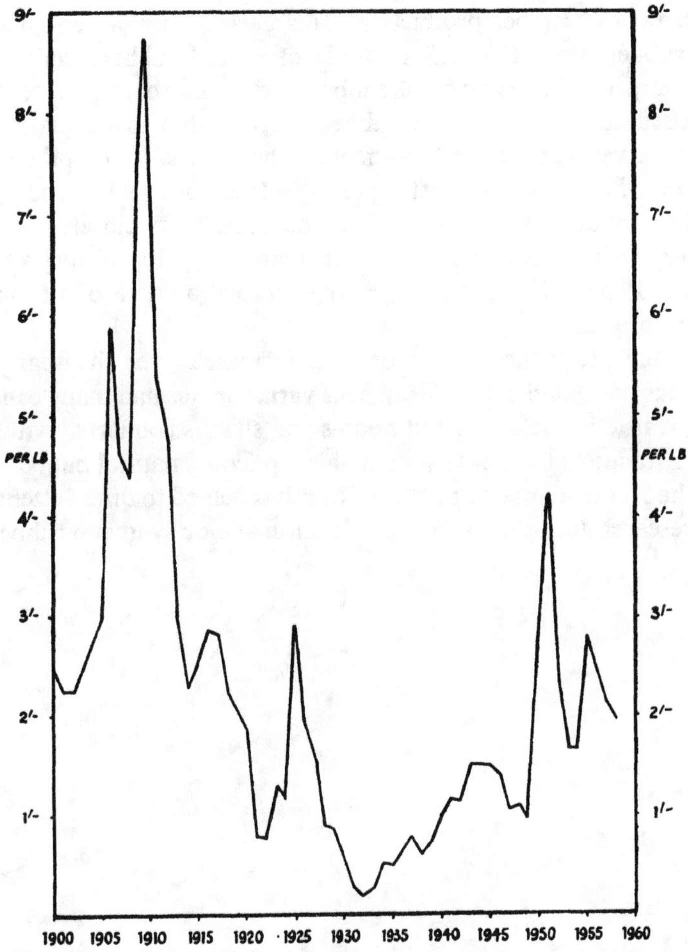

FIG. 4. Average price of natural rubber since 1900
The highest price in the year has usually been much higher

invention was forgotten. In 1888 J. B. Dunlop, a veterinary surgeon of Belfast, re-invented it, this time for his son's tricycle. The idea gradually gained favour, first for cycles and then for motor cars, as the added comfort, greater speed and ease of steering obtainable with pneumatic tyres became more obvious. In 1895 a motor car with pneumatic tyres was driven in the Bordeaux to Paris race. It can be safely said that the rapid growth of the motor-car industry would not have been possible without the pneumatic rubber tyre; and as rubber aided the growth of the motor industry the motor car assisted the growth of rubber production. This meant, very largely, plantation rubber, since this was the only organized rubber production. The sharp rise in the price of rubber during the years 1900 to 1910 increased the profits of the rubber growers and made plantation rubber a very attractive investment. The expansion of plantation rubber, which quickly outstripped all other forms of rubber production, followed the growth of the motor industry closely, and the history of natural rubber became in many ways the history of the pneumatic tyre. This is not surprising, since about half of all rubber goes into tyres.

As the production of rubber has increased over the years the price has oscillated a good deal. This variation has had many causes: wars, restriction schemes and booms and slumps connected with the motor trade; but the intermittent high price of natural rubber has itself had a most important effect, for it has helped to direct attention and research into the problems of man-made or synthetic rubber.

2

The nature of natural rubber

THE PHYSICAL PROPERTIES OF NATURAL RUBBER

ALL cooks know that if you warm sugar up in a pan it will soon melt to a clear liquid. Allowing the liquid to cool down produces a solid again, which looks and tastes the same as the original sugar—and of course it is the same. But if on melting the sugar the heating is carried just a little too far, the liquid will go brown and give the characteristic smell of burning sugar. It is now impossible to get the charred residue back into its original form because it has decomposed.

These two changes, the melting and the charring, are known respectively as physical and chemical changes. With the physical change the sugar can be returned to its original form fairly easily, but with the chemical change the result is no longer sugar. It is the same with physical and chemical properties. A physical property is the ability to undergo a change without losing chemical identity. A chemical property shows the ability of a substance to change to something else.

Several physical properties of natural rubber come fairly easily to mind. For example, if a piece of rubber is stretched its shape changes. This is a physical change, for it still remains rubber, and the property associated with it is the ability to stretch. To describe the physical properties of natural rubber, however, we must first say exactly what we mean by natural rubber, and this is not so easy.

To begin with, natural rubber can refer to either raw or vulcanized rubber, and it makes a difference which one is chosen because they do not have the same properties. Supposing we choose vulcanized rubber. If the vulcanization has been carried out with a little sulphur we shall obtain a tough elastic solid. If vulcanization is carried out with more sulphur and the heating is continued for a longer time, Ebonite is produced, which is hard and has no elasticity at all. In addition to this, incorporating other substances into rubber before vulcanization can alter the properties profoundly.

Two questions arise: what is to be considered as typical natural rubber, and what properties are to be associated with it? Probably the most common, and therefore the best example of natural rubber to take, is natural rubber vulcanized with a little sulphur to the tough and elastic state. This is usually known as 'soft' vulcanized rubber. On the question of properties, a compromise is usually reached between what seem the obvious physical properties and what can most easily be measured. The reason for this lies in the importance of being able to measure properties. It is essential to measure properties if different rubbers, or different states of vulcanization, are to be compared. For this reason it is sometimes better to consider a property in terms of others which can more easily be measured.

Elasticity, for example, is usually measured and studied physically by thinking of it in terms of a number of other properties. These are usually grouped under two headings: tensile properties, which deal with the strength of rubber, and dynamic properties, which deal with rubber in motion.

Tensile properties

A piece of high quality natural rubber, which has been vulcanized to a state producing the best possible stretching properties, will extend to 10 times its own length before breaking. Just before it breaks the rubber will carry a load equivalent to 10 tons per square inch. Apart from synthetic rubbers, and there are few which will approach these limits, no other material is known which will behave like this.

These figures are obtained, quite simply, by loading a piece of rubber with weights until it breaks, and measuring how far it has stretched in the process. The weight a sample requires to break it, expressed in pounds per square inch or other convenient units, is called the tensile strength.

This is a very widely used measurement for comparing the strength of rubbers. But it does not say much about what a particular rubber will be like in service, because very few rubbers are expected to be stretched until they break. Taking the weight which will stretch a rubber to a given elongation, say 100 per cent, is a

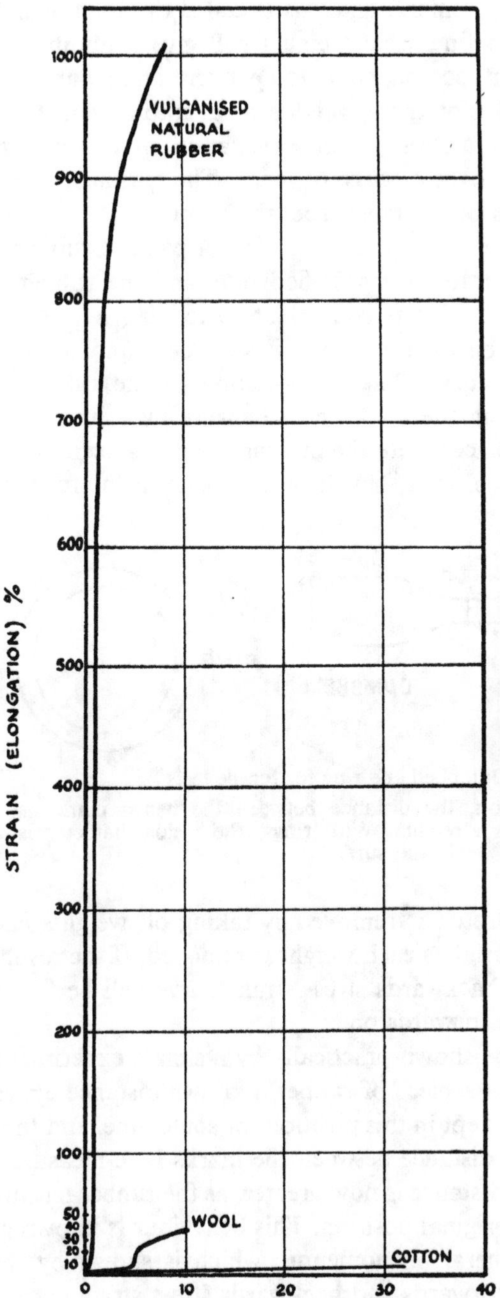

FIG. 5. Stress/strain behaviour of various materials

In order to show the very large difference between natural rubber and other materials, the results have all been shown on the same scale. In this instance the stress has been calculated on the actual cross-section of the rubber tensile piece when it broke. It is more usual to calculate it on the cross-section before stretching takes place. See Fig. 7 for a more normal-looking stress/strain curve for natural rubber

rather more useful measurement, and this is called the modulus. A better method still is to find what weights will give a number of elongations, 100 per cent, 200 per cent, 300 per cent and so on until the rubber breaks, and plot these weights as a graph against the elongation. This is called a stress-strain curve, and it gives a picture of what a rubber is like over a range of loads. The tensile strength and the moduli can also be obtained from the graph.

Tensile tests are made on a special machine. A piece of rubber is held in jaws which separate at a specified rate, and the rubber is stretched until it breaks. This is equivalent to loading a piece of rubber with weights. The load at break is recorded, and modern machines also draw a curve. The cross-section of the rubber is measured before applying the load, and to ensure that the rubber actually breaks at the place where the measurement has been made, the pieces are either dumb-bell shaped or are made in the form of rings.

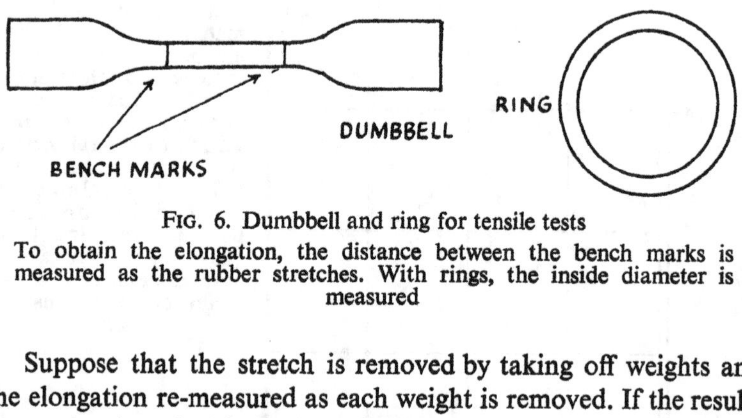

FIG. 6. Dumbbell and ring for tensile tests

To obtain the elongation, the distance between the bench marks is measured as the rubber stretches. With rings, the inside diameter is measured

Suppose that the stretch is removed by taking off weights and the elongation re-measured as each weight is removed. If the results are again plotted, this backwards stress-strain curve will be found to be different from the forwards one.

The difference can be shown practically by a simple experiment. Two marks are made on a piece of rubber a known distance apart. The rubber is stretched, kept in this position for some time, and then allowed to retract. The distance between the marks is re-measured, and it is found that the distance is now greater, as the rubber has not quite come back to its original position. This behaviour is known as permanent set. The general phenomenon, which is shown by the difference between the forwards and backwards stress-strain curves, is known as hysteresis.

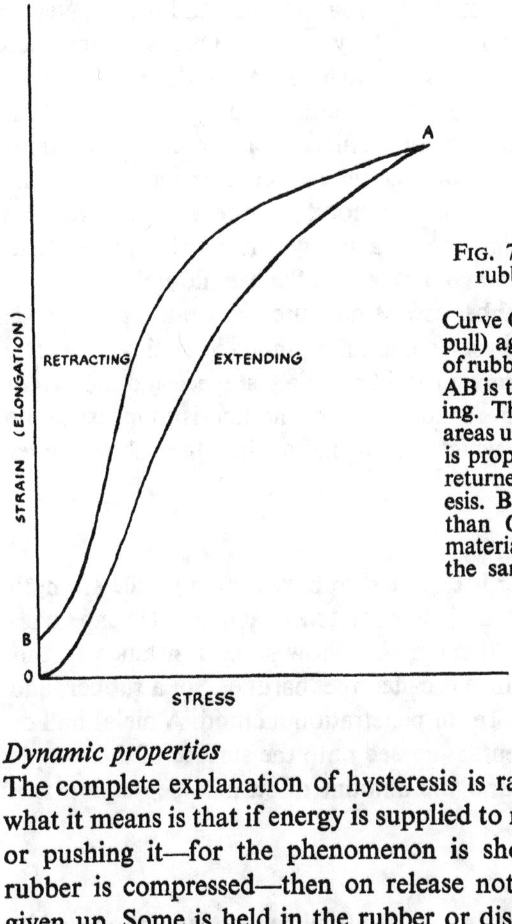

FIG. 7. Stress/strain curves for rubber, showing hysteresis

Curve OA is the plot of stress (i.e. pull) against strain, while a piece of rubber is being extended. Curve AB is the same while it is retracting. The difference between the areas under the curves, area OAB, is proportional to the energy not returned and is called the hysteresis. B is at a higher elongation than O, which shows that the material does not come back to the same place. OB is therefore the permanent set

Dynamic properties

The complete explanation of hysteresis is rather involved. Roughly what it means is that if energy is supplied to rubber, either by pulling or pushing it—for the phenomenon is shown equally well when rubber is compressed—then on release not quite all the energy is given up. Some is held in the rubber or dissipated as heat, and the overall effect is that the rubber does not quite return to its original dimensions.

The hysteresis that a rubber shows determines another very important property—its resilience or property of springing back after it has been under tension. Resilience is studied under the heading of 'Dynamic properties', not only because the springing back may be thought of as rubber in motion, but also because the resilience of a rubber is very important when the rubber is in continuous movement.

A tyre, for example, must have good resilience or its cushioning effect will be small, but it must also be resilient for another reason. A tyre on a moving vehicle is continually being compressed and released as each part of it meets the road surface and then leaves it. The compressions are small, but they are continuous. After each compression the tyre has a small permanent set, in this case called a compression set, and the energy not returned may be used to heat up the tyre. A rubber with poor resilience will hold back a larger amount of energy than one with better resilience and as a result the tyre will get hotter. If a tyre gets too hot it will break up, and for this reason the tread on a high-speed racing tyre is much thinner than that on a normal tyre, to avoid excessive 'heat build-up'.

The resilience of a rubber can be measured by drawing hysteresis curves, but this is not a very common method. Sometimes a metal ball is dropped from a known height on to a test piece and the height of the rebound measured. A more modern method is to measure the rebound of a pendulum which has swung against the rubber under test.

Hardness

Although a useful rubber is expected to have a high tensile strength and good modulus, that is, it is expected to change its shape considerably under loading, it must also show some resistance to this deformation. This resistance is called the 'hardness' of a rubber, and it is measured by some form of penetration method. A metal ball or a blunt-pointed instrument is pressed onto the surface of the sample under a constant load, and the amount of surface depression is a measure of the hardness.

Abrasion

The abrasion, or the wearing away of rubber, shows up one of its most remarkable properties. Since it is an elastic substance and can 'give' under pressure, it can show a resistance to abrasion that is a great deal better than that of steel.

Once again, however, abrasion is a composite property, and accurate measurement of it difficult. The abrasion resistance is usually measured by bringing a piece of rubber into contact with a wheel or drum revolving at high speed. But most laboratory abrasion-testing machines give different results for different types of machine. Since one of the chief reasons for measuring abrasion is to find out how a rubber will behave in service an alternative method is to try the rubber out under service conditions. This is an expensive and not a very satisfactory solution, but in the case of tyres extensive road-testing over all types of surface is a very necessary alternative.

Abrasion is also one of the examples of ways in which rubber properties can be altered by changes in the composition of the final product. The high resistance to abrasion which rubber can show has already been mentioned. Suitable alteration of the composition can give a product which wears away more easily than paper—for example, in pencil erasers.

Electrical properties

Normally rubber offers a good deal of resistance to the passage of electricity, and is therefore a good insulator. But here, as with many other properties, adding different substances to rubber can substantially alter the property. Carbon black, more generally known as soot, normally gives a mixture with rubber which is an insulator, although carbon black is itself a conductor of electricity. But by using special types of black the mixture can, in fact, be made to conduct electricity.

Although rubber is a good insulator it tends to absorb small amounts of water, and this breaks down the insulation. Natural rubber is therefore of no use in making submarine cables. A very near chemical relative of natural rubber, known as gutta percha, has been much more successful, as it absorbs much less water, and it was used for this purpose for nearly a hundred years. It is now, however, being replaced by plastics. Ebonite, the hard form of vulcanized rubber, has also much less water absorption and makes a good insulator in switch mountings for example.

THE CHEMISTRY OF NATURAL RUBBER

Atoms and molecules

Atoms are so small that the distance across a pin-head, about a millimetre, will accommodate 10,000,000 average-sized atoms placed side by side. Think of the number there must be in the whole pin-head, or in a piece of coal, or a mountain; and considering the tremendous number of atoms there must be in the whole world, it is rather surprising that there are only about 100 different chemical kinds altogether—rather less than the number of different kinds of birds found in and around the British Isles.

Substances containing atoms of one kind only are called elements. Some examples of elements are the gases oxygen and nitrogen which largely make up the air we breathe, hydrogen the lightest element known and carbon of which coal is made. There are in fact about 100 elements corresponding to the 100 different kinds of atoms.

The most interesting chemical property of atoms is their ability to combine or link up with each other to form compound atoms or molecules, as they are usually called. This is the same thing as saying that elements react or combine to form compounds. For as the atom is the smallest unit of an element, the molecule is the smallest unit of a compound.

When atoms combine in this way they form quite definite bonds or links with each other, and since one atom may sometimes combine with several others at the same time, quite complicated patterns can be built up. We shall meet an example of this later on.

A simpler example is the water molecule. The elements hydrogen and oxygen can be made to react to give water, and when they do two atoms of hydrogen combine with one atom of oxygen to give the water molecule, which contains three atoms. This can be expressed by writing the formula, which is a kind of short-hand, H_2O. Another way of looking at it is to say that two parts of hydrogen combine with one part of oxygen, which is what we should find if we analysed water. Analysing a compound into its elements, however, only gives the proportions of the elements combined, it does not always give the number of atoms of each type in the molecule.

The formula of natural rubber

We have seen that the formula of a compound gives the proportions of the different elements which make it up, or the ratio of the number of atoms contained in the molecule—which is the same thing.

The formula can often be obtained by analysis. One of the first successful analyses of natural rubber was carried out by Michael Faraday in about 1826, on a sample given him by Hancock. The analysis was by no means easy as the dried rubber used by Hancock and others was a complex mixture of chemical substances. But Faraday carefully separated off the rubbery substance and on analysis found that it contained two elements, hydrogen and carbon, in the ratio of eight parts of hydrogen to five parts of carbon.

We might assume, to start with, that the rubber molecule contains eight atoms of hydrogen and five of carbon, and that the formula might be written C_5H_8. How are the atoms arranged? Chemists know that the hydrogen atom can combine with only one other atom at a time, but carbon can combine with four, three or two other atoms at the same time. There are therefore a large number of ways in which eight atoms of hydrogen can be linked with five atoms of carbon, and only one particular arrangement will represent natural rubber, all the others will be different compounds.

To decide how the atoms are arranged in a molecule, chemists usually try less destructive forms of analysis than that used by Faraday. In other words they try to break the compound down, not into its elements, but into simpler compounds in which the atomic arrangement is known. Then by studying the products of analysis they try to work out how the original molecule was put together.

This is not a very easy thing to do, even today with all the refined techniques of analytical chemistry available. In the middle of the last century, when this problem was first tackled, there were far fewer methods available for analysis and the problem was proportionately more difficult. It is all the more to the credit of the early pioneers that they made the progress they did.

One of the first things they tried was the effect of heat on rubber. This gave rise to a great many substances, most of which could not be identified at the time, but careful work showed that one substance was nearly always produced, no matter how the heating was done. It was not long before chemists were saying that this was the important derivative of natural rubber. Unfortunately the arrangement of the atoms in this molecule was also a mystery, and it was not until 1884 that the substance—which is called isoprene—was finally analysed to everyone's satisfaction.

The strange thing is that isoprene has exactly the same formula as natural rubber (C_5H_8). Isoprene and rubber are obviously not the same compound, since isoprene is a colourless liquid and rubber a brown elastic solid. The answer seemed to be that whereas isoprene has only eight atoms of hydrogen and five of carbon, rubber has many more than this but the hydrogen and carbon atoms are in the ratio of eight to five. At first sight this made the problem look even more difficult, but since by heating natural rubber you could obtain isoprene, chemists thought that the molecule of natural rubber might be a string of isoprene molecules joined end to end.

This, as it turns out, is not so far from the truth. Natural rubber does have a very large molecule, and it is made up of repeating units. But the units, though they are very similar to isoprene molecules, are not quite the same. The formulae of isoprene and part of a natural rubber molecule are given together for comparison in Fig. 8.

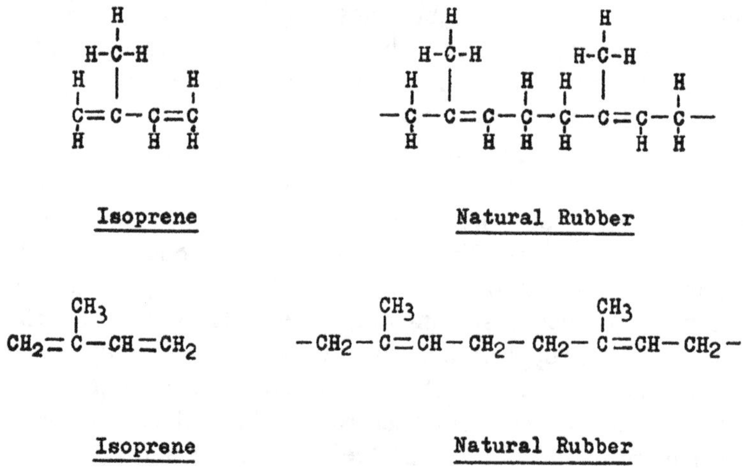

FIG. 8. (The bottom formulae are merely shorthand ways of writing the top)

It can be seen that isoprene, written as in the diagram, has a chain of four carbon atoms, while one carbon atom sticks out of the chain. Two of the linkages or bonds joining carbon atoms are represented by double lines. This is merely a different type of bond from the one represented by a single line.

Comparing isoprene and rubber it can be seen that in rubber each isoprene unit has only one double bond. Natural rubber molecules vary in length, and contain normally between 1,000 and 40,000 repeating units, so that they still contain a great many double bonds. It is these double bonds which give rubber a number of distinctive properties. It might be thought that a double bond would be twice as strong as a single bond. But in fact the reverse is true, for double bonds show a marked tendency to change into single bonds and this means that a substance containing double bonds is much more reactive chemically than one with only single bonds.

The double bonds make rubber susceptible to attack by oxygen in the air. This is why thin films of raw rubber go sticky and eventually decompose when left for long periods in the open air. But while the double bonds allow rubber to be attacked and broken down by oxygen they also enable it to be vulcanized. If rubber and sulphur are heated together, instead of breaking the rubber molecules the sulphur atoms attach themselves to the chains and link them together. This is vulcanization, which results in a decided improvement of the rubber. In fact, double bonds are so important for vulcanization purposes that some synthetic rubbers are deliberately made with double bonds to enable them to be vulcanized.

The elasticity of natural rubber

Ignore for a moment the division of physical properties into tensile strength, resilience, etc. The fundamental thing about rubber, either raw or vulcanized, is that it can be stretched to considerably more than its original length, and when released it will snap back to very nearly its original dimensions. This behaviour is called elasticity, and it can be explained by reference to the rubber molecules.

The important thing about these molecules is that left to themselves they tend to be coiled up, not in any regular fashion, but unevenly twisted and twined, very much like a plate of spaghetti or a tangled skein of wool, except that a skein of wool will consist of just one strand and in the tangled mass of molecules there will be many. Fig. 9 shows the sort of shape a molecule of this type can assume. It can in fact assume almost any shape, but the mathematicians tell us that it is much more likely to be coiled up than uncoiled. The molecule, however, need not always be coiled. If enough force is applied at its ends it will unwind. The force can be applied by pulling the ends of a piece of rubber, and since the distance between the ends of a molecule is greater when it is straight than when it is coiled up the rubber stretches.

The question now arises, why doesn't a piece of string or a steel wire stretch as well? The answer is that they do to a limited extent. They do not stretch as much as rubber for a number of reasons, the most important of which involves the movement of the molecules. All molecules move about if they can, and atoms too for that matter, the movement being connected with temperature. The hotter a substance, the more its atoms and molecules move about. But

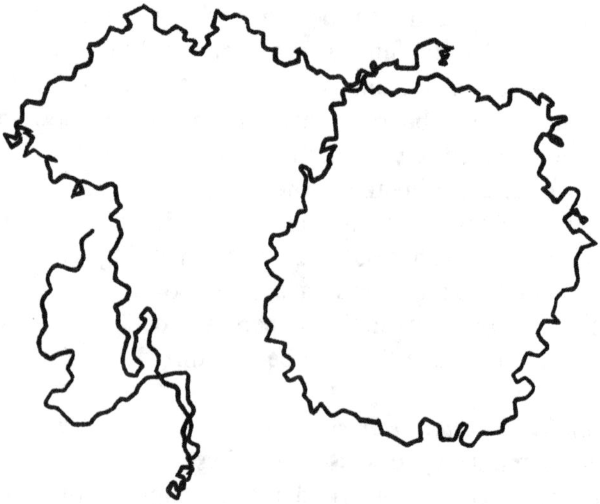

FIG. 9. A representation of part of a rubber molecule

movement is restricted if the atoms and molecules are packed closely together or if they strongly attract each other as some do.

The atoms in a piece of steel are packed closely together and movement is very restricted. This situation is common to most metals, and stretching of the kind shown by rubber is not usually associated with them.

A piece of string or a strand of wool are rather different. While it cannot be said that a piece of metal has much in common with rubber, the natural fibres—cotton, wool and hemp—are quite similar. They have long chain-like molecules, for one thing. But they still do not stretch much. The reason is that the molecules are not tangled up like rubber molecules but are arranged in a more or less orderly manner. A fair amount of attraction exists between the molecules and little movement is possible.

For good stretching a long flexible molecule is required, with not much attraction between the chains. Some attraction between molecules is necessary because otherwise the chains would slip past each other when pulled, and when the pulling stopped they would have no tendency to go back to their original positions. The substance would then be plastic or dough-like. This can be avoided by having a long molecule which is more likely to be well tangled than a short one, and well-tangled molecules will lie close to each other at a few points—the knotty part of a tangled skein of wool—and far away at others. Now all molecules attract each other to a certain extent, the attraction being greater the nearer they are together. Where the molecules are tangled close together the attraction will be great and the movement small. This will enable the molecules to come back to something like their original positions, and the rubber will retract when pulling stops.

The stretching of natural rubber is not as good as it might be. The elasticity of raw rubber can be markedly improved by vulcanization with sulphur, for example. Vulcanization with sulphur leads to the formation of sulphur atom bridges between adjacent molecules. Only a few of these bridges are formed for each molecule, but it is enough to reinforce the natural attraction, and therefore to improve the ease of retraction and thus the elasticity.

It seems surprising that natural rubber is the only example found in nature of long-chain molecules made up of carbon and hydrogen atoms and having elastic properties, for this type of molecule is not difficult to make in the laboratory. As we shall see later, a great many have been made in the last 30 years, nearly all with wide commercial applications, for these are the synthetic or man-made rubbers.

PART TWO

Synthetic Rubber

3

Historical introduction to synthetic rubber

THE English chemist Greville Williams was one of those early pioneers who succeeded in obtaining isoprene by heating natural rubber, and who guessed that isoprene was its parent substance. In 1860 he carefully purified some isoprene, and after allowing it to stand for some time, heated it and obtained 'a white spongy elastic mass'. This has been claimed—though not by Greville Williams—as the first man-made or synthetic rubber. It is very doubtful if it was a synthetic rubber, as subsequent chemists repeating the experiment could find nothing like rubber in the result. A few years later a Frenchman, Bouchardat, made what was probably a much better approximation to natural rubber by heating isoprene with hydrochloric acid. But here again many people say that this was not a true rubber either.

The question of who may be credited with first having made a rubber in the laboratory has been argued with some bitterness for many years. In England, at least, the credit is usually given to the Birmingham chemist Tilden, who was also responsible for proving the structure of isoprene. Tilden's discovery, which he made in 1882, was apparently accidental. He had kept bottles of isoprene standing on his shelves for many years, and one day he noticed that large yellowish lumps were floating about in the liquid. Tilden says that the lumps were tough and elastic like natural rubber, and it is probable that this was the nearest approach to natural rubber which had been made up to that time.

The process of linking up small molecules like isoprene to form long chains was not a new one even in Tilden's time. Since the chains were made up of repeating units they were called polymers—from the Greek *poly*, meaning 'many'. The small molecule used to make the units in the chain was called, not unnaturally, a monomer, and the process referred to as polymerization. Tilden's reaction had apparently occurred of its own accord and he called it a 'spontaneous' polymerization.

Tilden was enthusiastic about his man-made rubber. He suggested that if an alternative source of isoprene could be found the production of synthetic rubber on a large scale might be economically possible. At that time it was an attractive idea. The only source of natural rubber was the wild rubber of Brazil and Africa, the quality being variable and the price high. Tilden himself later discovered an alternative source of isoprene: turpentine. But since turpentine could only be obtained from trees it suffered the same economic disadvantages as natural rubber. It was not cheap, its quality was variable, and it was subject to market manipulations in the same way as natural rubber.

It seemed that Tilden's high hopes for a synthetic rubber industry were not to be realized. Apart from the fact that his synthetic rubber was a very poor substitute for natural rubber, he reported in 1900 that the polymerization was still taking a long time, and he could find nothing to speed it up. Everything he tried yielded a sticky oil which was quite useless. But in the same year Kondakov, a Russian chemist, reported that dimethyl butadiene, a near-relative of isoprene, could be converted into an elastic mass by heating it with potash. Once again the product did not compare very favourably with natural rubber, but it was an important step forward because for the first time a different chemical substance with elastic properties had been prepared.

Kondakov has been hailed as yet another chemist who was the first to produce synthetic rubber. This claim is not as unreasonable as it sounds, for today a synthetic rubber is almost invariably a substance which has been made from a chemical other than isoprene.

This use of the word 'synthetic' has been frowned upon by many chemists, as, strictly speaking, 'synthetic' is a technical term and has quite a definite meaning. A synthetic substance is one which has been built up or synthesized in the laboratory, the qualification being that the substance must have exactly the same composition and structure as the natural product. For example, certain dyes like indigo which are found in plants have been synthesized, and synthetic indigo is identical to the natural product. Until recently it could be said that natural rubber had never been satisfactorily synthesized. None of the so-called synthetic rubbers was a laboratory copy of natural rubber, they were quite different substances, and they hardly warranted the name of rubber. But continued use has made the term acceptable, and today any material which stretches to twice its length, retracts when released and can also be vulcanized is called a rubber, and to distinguish the man-made products from the natural it is useful to call them synthetic. Unfortunately the misappropriation of the word 'synthetic' has had the following result: when a reasonable laboratory copy of natural rubber was made recently the only name left was 'synthetic natural rubber'.

The beginnings of synthetic rubber production

Although Tilden became pessimistic about synthetic rubber, and prophesied in 1900 that it would never be a practical proposition, the early years of the twentieth century witnessed spectacular progress in the subject. Kondakov's discovery that dimethyl butadiene could be made to give a rubbery polymer was followed in 1910 by the discovery of another Russian chemist, Lebedev, that butadiene could be polymerized. Butadiene may be looked upon as the parent substance of both dimethyl butadiene and isoprene, and the surprising thing is that no one seems to have tried to polymerize it before. Butadiene, however, is a gas at room temperature, and until good refrigeration facilities became available in laboratories it was difficult to liquefy, and therefore difficult to handle.

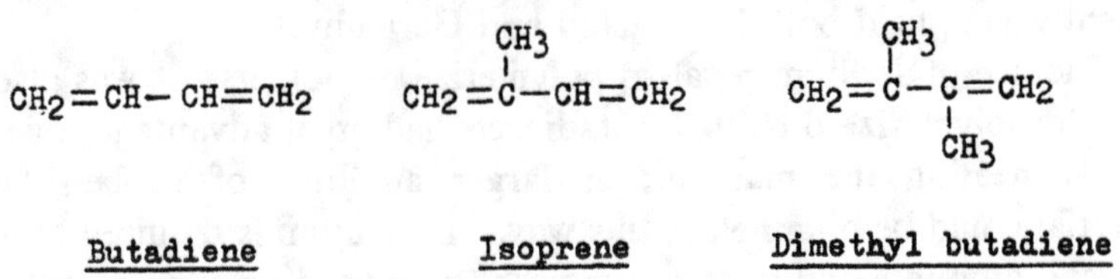

FIG. 10. Formulae of butadiene, isoprene and dimethyl butadiene

The intense activity in the field of synthetic rubber which took place in the early years of this century was not due to academic interest alone. The demand for rubber was steadily increasing with the growth of the motor industry, but the production of natural rubber (wild rubber) was not keeping pace because of over-exploitation of trees and native populations. There was no other substantial source of natural rubber as the export of plantation rubber had not reached any appreciable volume. All these factors combined to push up the price of natural rubber, and during 1910 it reached the all-time record of 15*s*. per lb. An alternative source of rubber was an attractive proposition.

In Germany much work was going on in this direction. A source of isoprene other than turpentine was found, and a rubbery substance was made by heating isoprene in a sealed tube for about eight days. This was a marked improvement on Tilden's spontaneous polymerization, which took several months, but the rubber was still of very poor quality compared with natural rubber. Work was also begun on the polymerization of dimethyl butadiene and of butadiene, and many sources of these were investigated. An interesting suggestion made in Germany at this time was that both isoprene and butadiene might be obtained from petroleum. It is only in recent years that this idea has been developed, and today petroleum is the chief source of these monomers.

In England a team of distinguished chemists, which included Sir William Ramsey and William Perkin (whose father discovered the first coal-tar dye), were also looking for cheap sources of the three monomers. An alternative route to isoprene was worked out and in 1910 two chemists—Matthews and Strange—found a much quicker way of polymerizing isoprene, by using the reactive metal sodium. They applied for a patent on October 25th of that year, and three days later the German chemist Harries announced the same discovery. A legal battle took place, but the Matthews and Strange patent was upheld both in England and Germany.

The use of sodium metal to polymerize monomers—it was also used to polymerize dimethyl butadiene—had great advantages over merely heating the material, as larger amounts of rubber-like material could be obtained in this way. The sodium is required only in small amounts and does not appear to be used up very quickly. It is therefore called an initiator or sometimes a catalyst.

Work was also going on in Russia and the United States. In Russia a number of routes to the three monomers were worked out, and sodium polymerization of butadiene begun. This reaction was to remain for many years the staple source of synthetic rubber in Russia.

Synthetic rubber in the First World War

In July 1916 a German submarine docked at the American port of Baltimore and loaded 500 tons of natural rubber for the return journey across the Atlantic. The same submarine made a further trip to the United States and carried a similar return cargo. On the third attempt she was sunk. Germany was at war, under blockade and virtually without natural rubber. England and France, with a more extensive command of the seas, had ample supplies from Malaya, where plantation rubber was by this time in production in large quantities. Germany started the war under the hopeful impression that supplies of natural rubber would be fairly easily obtained. But after a year of blockade, and when it became obvious that very little could be obtained by smuggling, attention was turned to the problem of making synthetic rubber.

After much discussion it was decided to make synthetic rubber by polymerizing dimethyl butadiene, and this was to be obtained from acetone, the acetone in turn being made by the fermentation of potatoes. But no potatoes could be spared as they were required for food, so the acetone was made from calcium carbide, the substance well known to users of early cycle lamps.

Two methods of polymerizing dimethyl butadiene were used giving two types of synthetic rubber, designated by the letters H, from the German word *hart*, meaning 'hard', and W from the word *weich*, meaning 'soft'. The H rubber was made by allowing dimethyl butadiene to stand in drums, not quite full, for about three months. This was called 'cold' polymerization, and the resulting polymer, which gradually filled the container, was white and puffy-looking rather like a cauliflower. The W rubber was made by heating the monomer in drums for about six months, and the rubber was yellow, transparent and tough.

Compared with natural rubber both the synthetics were very poor indeed. The H rubber gave, when vulcanized, products which were less than one-third as strong and only one-third as extensible as natural rubber. The W rubber gave vulcanized products nearly as extensible as natural rubber but only one-tenth as strong. Nevertheless, the H rubber was vulcanized to the hard state and used for accumulator boxes, aerials and discs for magnetos; and the W rubber was pressed into service for making vulcanized balloon fabrics and insulation for wires. Solid tyres were also made of synthetic rubber, but they lacked elasticity, and pieces could be chipped off them in cold weather, while in hot weather the car had to be jacked up to prevent the rubber running into flat spots. Later, a third type of synthetic rubber was made. This was called B rubber and was produced by allowing dimethyl butadiene to stand in contact with sodium, in an atmosphere of carbon dioxide. This proved to be much better than either H or W rubbers for covering wires, though the vulcanized products it gave were no stronger than either of the others.

Improvements were being made though, and by the end of the war Germany was making synthetic rubber at the rate of 150 tons a month, and additional plants were in the course of construction with a combined capacity of 8,000 tons a year.

Progress between the wars

When the First World War came to an end work on synthetic rubber abruptly ceased. German manufacturers were glad to get back to natural rubber, after the very poor quality rubbers they had been forced to use. The price of natural rubber was down to 1*s*. 6*d*. per lb. and no one thought much about synthetic rubber.

This state of affairs lasted into the 1920's, but interest awakened in Germany in 1925 when the price of natural rubber shot up to 6*s*. per lb. A little work was then being done in the United States, by some immigrant Russian chemists, and the subject was being revived in Russia itself. It must be remembered that the plantations in the Far East were largely British and Dutch owned. Synthetic rubber was therefore a subject of recurring interest to those countries with growing motor industries and no control over the price of an important raw material.

By 1926 work on synthetic rubber was being conducted in Germany by I. G. Farbenindustrie, the large chemical combine. Of the three monomers, dimethyl butadiene, isoprene and butadiene, dimethyl butadiene was dropped, as experience in the war had shown that its polymers were of poor quality. The rubbers obtained from isoprene and butadiene were better, but isoprene was not so easily or cheaply manufactured as butadiene, and work went on therefore on the polymerization of butadiene.

The polymerization was still being conducted using sodium as an initiator, and the series of rubbers produced by making small changes in the formula was given the generic name Buna, with numbers to denote the type of rubber. Buna is derived from the first two letters of the words butadiene and natrium, which is the German name for sodium. The numbered Bunas were not very good as substitutes for natural rubber, but some were adequate in the hard rubber field, which recalls the moderate success of H rubber for accumulator cases during the First World War.

Soon another method of polymerizing butadiene was being tried. This was called emulsion polymerization and it involved polymerizing the monomer in a liquid, usually water. Since butadiene will not mix with water, even compressed and cooled to liquefy it, but floats on the top, some sort of soap was required as a mixing agent. The result was a mixture of small globules of liquid butadiene dispersed in the water, in other words an emulsion. The original intention was to reproduce the conditions of formation of natural rubber, which appears in the rubber tree as an emulsion in water: latex. Patents had been taken out for the idea as early as 1912. Since sodium metal reacts violently with water it could not be used as an initiator, and others had to be found. The product, however, was still called Buna rubber. Emulsion polymerization proved to have many advantages over any other method then in use. It was faster and it gave a more complete reaction. The product was a better rubber too, though still not as good as natural rubber.

The problem of the inferiority of synthetic rubber as compared with natural rubber was a serious one. For example, butadiene-polymerized rubber was very difficult to process. It was hard and tough and not easy to masticate. More time and power had to be used, which made it expensive. To overcome this it was decided to try to mix different rubbers together to see if a combination of properties could be obtained. For example, a softer rubber might confer a little more plasticity on butadiene rubber. The mixing of rubbers, however, gave disappointing results.

An extension of this idea was to polymerize different monomers together. This was not a difficult thing to do as monomers often polymerized under the same conditions, with the same initiator. The process gave much better results than merely mixing rubbers together, and it was extended to include many other monomers known at that time. One of these was styrene, a clear volatile liquid. Styrene can be polymerized itself to yield, not a rubber, but a hard glass-like solid called polystyrene, which is not elastic and cannot be vulcanized. It is in fact a plastic and is often used today for making ashtrays and ornaments. The important thing is that polystyrene is a thermoplastic: it softens or becomes plastic when heated. Since butadiene rubbers lacked plasticity it was not long before butadiene was being polymerized with styrene, and the resulting polymer—called a co-polymer—was very encouraging. It was much more easily processed than the original butadiene rubber and it had better physical properties. The rubber was called Buna S, the S standing for styrene.

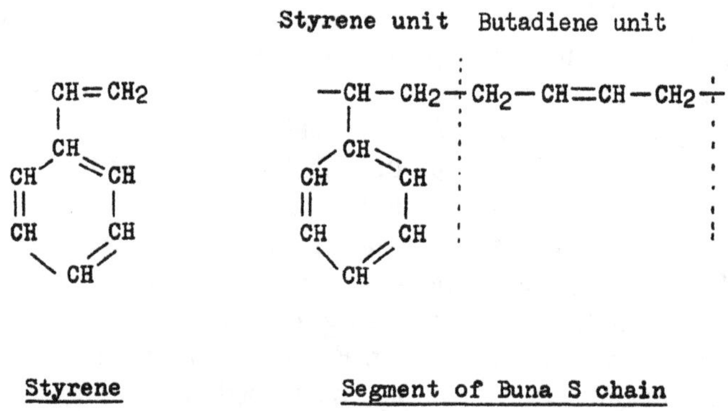

FIG. 11. The six-membered ring structure in styrene is stable and the double bonds in it are not as reactive as those in the chain. For the purpose of polymerization styrene has only one effective double bond, the one not in the ring structure

The field of polymerization was steadily widening. Chemists were looking around for anything that was likely to polymerize, and about this time another success was achieved by polymerizing butadiene with another monomer called acrylonitrile. Acrylonitrile was a new departure in rubber monomers. It contained not only carbon and hydrogen, as all the others had done, but nitrogen as well.

As might be expected, the properties it confers on the rubber—a co-polymer of butadiene and acrylonitrile—are somewhat different from the normal run of rubbers. It gives a rubber which has a remarkable resistance to attack by oils, a feature which is very desirable in some applications, as even the best vulcanized natural rubbers if immersed in oil for any length of time suffer swelling and slow decomposition.

This rubber was called Buna N, and it was one of the first of the so-called special purpose rubbers. They were rubbers with little similarity to natural rubber, and in most of the rubbery properties

Part of a natural rubber plantation

Part of a synthetic rubber plant, showing the bottom half of a reactor

A Bridge-Banbury Internal Mixer mounted over a $22\frac{3}{16}''$ and $26'' \times 84''$ Two-Roll Mill

$24'' \times 68''$ Three-Roll Calender, equipped with flood lubrication and motorized nip control

they were inferior. But because they had outstanding properties not shown by natural rubber, for example a resistance to oils or being a good deal less inflammable, they were in constant demand. The price was usually much higher than that of natural rubber but they found a ready market. In contrast to the special purpose rubbers, rubbers made from butadiene and styrene, and sometimes isoprene, are usually thought of as general purpose rubbers.

The American contribution

The story of synthetic rubber in the Second World War is one of the most remarkable of modern times. It began in 1929 when the Standard Oil Company in America made an agreement to use the I. G. Farbenindustrie process for the hydrogenation of coal and oil. The Standard Oil Company was allowed access to the scientific work which was then being done in Germany on the process, and later an agreement was reached between the two companies whereby the Standard Oil Company manufactured butadiene from oil. It proved an expensive process, however, and was discontinued, but it began a patent agreement between the Standard Oil Company and I. G. Farbenindustrie which was to prove very useful to America in the forthcoming war.

In 1937 an explosion in an American Neoprene plant virtually stopped production of this special purpose rubber, and the makers appealed to the I. G. Company for some of their oil-resistant Buna N. This was soon forthcoming and proved so successful that the German makers decided to try to sell some of their Buna S, the butadiene-styrene rubber. German experts went to America with impressive performance figures, and later large quantities of the rubber were sent over for testing. In addition, early in 1939, I. G. sent an expert to America to advise on processing, compounding and vulcanization of Buna S. The interest of American manufacturers was steadily increasing and they undertook a certain amount of research themselves. Had the war not begun in September 1939 the German producers might well have sold considerable quantities of Buna S to America. But when the war began the German Government—mindful of the acute shortage suffered in the First World War—cut off all exports of synthetic rubber and began stockpiling natural rubber as well.

Britain began the war with no shortage of natural rubber and no restriction was laid on its use until 1940. In that year an agreement was made with the United States whereby Britain acquired a stock of American cotton and the United States began a stockpile of natural rubber. Rubber was far more of a strategic material in the Second World War than it was in the First, for in the Second armed forces were more mechanized and nearly all fighting machines, from aircraft to battleships, were dependent to some degree on rubber.

The reciprocal agreement between Britain and the United States proved to be a wise move, for when Malaya fell to the Japanese a large proportion of the natural rubber supply was cut off, and the stock-pile was sorely needed. The stockpile of natural rubber itself would not have been enough but this was augmented by synthetic rubber.

When the war in Europe began, Standard Oil started to build a plant for making butadiene and one for producing Buna N. Licences were offered to rubber companies to produce Buna S under the Standard–I. G. patents, and by 1941 four American rubber companies had begun to make synthetic rubber. Rubbers were produced under different names and different specifications. But a series of Government committees were appointed, patents were pooled, and the responsibility for production passed into the hands of the United States War Production Board. The Government financed the whole project. The types of rubber to be made were limited, and each was to be made to a standard specification. Eighty-seven factories were to be built, most of them to a standard design, and by 1942 plans were made for an annual output of 845,000 tons of synthetic rubber. By August 1944 all the factories had been built and the annual production was 1,000,000 tons of synthetic rubber. It had indeed been a race against time, for by the beginning of 1943 stocks of rubber were falling fast, and they continued to fall until the beginning of 1944, when production and imports just began to overtake the requirements.

By any standards the production of synthetic rubber in the United States was a fantastic achievement, and great credit is due to American industry and organization. But the ironic thing is that the main production of synthetic rubber was based on a co-polymer of butadiene and styrene, which was a German invention.

Developments after the Second World War

After the impetus given it by the last war, synthetic rubber has bounded ahead, and developments have appeared with a speed reminiscent of the early period of intensive development. When the war ended the United States decided to keep a certain amount of synthetic rubber production going as a nucleus for any future expansion occasioned by a national emergency, and as an alternative source of rubber in an effort to stabilize prices. The Government-owned factories were sold to the rubber and oil companies who had run them, and extensive research was kept going to try to improve the polymer.

The production of the American synthetic rubber was carried out by emulsion polymerization, as had been the Buna S on which it was based. But improvements in America during the war had led to a cutting down of the reaction time; and whereas Buna S required 25 to 30 hours at a temperature of 45° to 50° Centigrade, the American rubber required only 12 hours reaction time at 50° Centigrade. A new development concerning the initiator of the polymerization process was discovered independently in the United States, Britain and Germany. Only the German invention proved suitable for synthetic rubber, and it showed far more promising results in producing the American type of synthetic rubber than it did the German. When information became available about it after the war it was applied to synthetic rubber production in America.

By the use of the new initiator the temperature used for reaction was lowered from 50° to 5° Centigrade, that is to just above the freezing point of water, and the polymerization took no longer. The result, moreover, was a much better rubber, popularly known as 'cold rubber'.

Since then even more spectacular developments have occurred in the field of synthetic rubber, but these will be saved for a later chapter. Although at the present time changes and new ideas are still coming thick and fast, most of the world-wide production is at present based, and will be for some years to come, on butadiene-styrene rubber, and this will be the chief general purpose synthetic rubber described in this book.

4

The manufacture of general purpose synthetic rubber

THE difficulty in describing any manufacturing process is to know where to begin. In the Chemical Industry to speak of raw materials is to be rather vague, for the raw materials of a particular manufacturing process may be derived from yet other raw materials, and so on until one goes back to the real raw materials like coal, limestone and oil. The same is true of synthetic rubber. This chapter will be devoted to the manufacture of general purpose synthetic rubber made from butadiene and styrene, which might be considered its raw materials. But we can go even further back and discover how butadiene and styrene are made, and this seems to be the best approach because the economy of butadiene-styrene rubber is highly dependent on the ultimate materials from which butadiene and styrene are made.

The manufacture of this rubber will therefore be divided into two parts: the manufacture of the monomers, butadiene and styrene, and their co-polymerization to form synthetic rubber. This division occurs in practice because the monomers are often made in separate chemical plants and delivered to the synthetic rubber factories, by pipeline if they are near enough, or by tank car if they are not, and there the polymerization takes place.

BUTADIENE

Since one of the major aims of rubber chemists has been to produce a cheap rubber, one of the most important things about a particular way of manufacturing a monomer is its cost. Over the years many ways of making butadiene have been devised, and about each way these questions—among others—had to be answered: are the raw materials readily available? are they cheap? Since the cost and availability of the raw materials changes from time to time, the methods of producing butadiene have also changed.

Before the last war butadiene was made in Germany from acetylene, which in turn was made from calcium carbide and water. The same reaction was used to make acetylene in old-fashioned bicycle lamps. The raw materials for calcium carbide are limestone and coal, of which there were abundant supplies in Germany. So it might be said that synthetic rubber was made from coal, limestone and water. The route has now been abandoned because cheaper sources of butadiene are available. Nevertheless, acetylene continues to be a very important chemical, serving as an intermediate in many other chemical processes.

Another important chemical intermediate is alcohol, and butadiene has also been made from this, notably in America during the last war and also in Russia. The raw materials for alcohol include almost anything fermentable, but in Russia potatoes were mainly used, and in America wheat and maize. The alcohol route to butadiene has often looked an attractive one, particularly to farmers, and some time ago a bill went before the United States Senate with proposals for converting many millions of bushels of surplus corn (wheat) to butadiene via alcohol. The United States administration pointed out, however, that the production of butadiene from alcohol cannot compete in cost with its production from petroleum or natural gases. This is very largely the position at the moment. Nearly all the world production of butadiene is based on petroleum.

Petroleum

Petroleum is usually found in underground deposits, and sometimes even beneath the sea-bed. It was first considered an intolerable nuisance when the oil seeped into salt mines and contaminated the salt. Later it was sold, often mixed with salt, as a 'cure all', and-later still it was used in lamps, replacing whale oil. Nowadays it feeds the motor industry, synthetic rubber and plastics and many other chemical industries.

This is largely because crude petroleum contains a seemingly inexhaustible number of different chemical compounds. Like isoprene they are nearly all hydrocarbons, that is to say they contain the elements hydrogen and carbon only. The complete separation of these hydrocarbons is very difficult indeed, but a partial separation may be achieved by distillation. If the oil is heated the lower boiling hydrocarbons boil off first, and if the vapours are condensed and collected at intervals the result is a series of fractions which boil at a steadily increasing temperature.

The first fraction is the one from which petrol or motor spirit is made. The next is called the paraffin-oil fraction, and this is used in heating and lighting and in certain jet-propulsion mechanisms. Still higher boiling fractions are the source of lubricating oils and greases, such as vaseline.

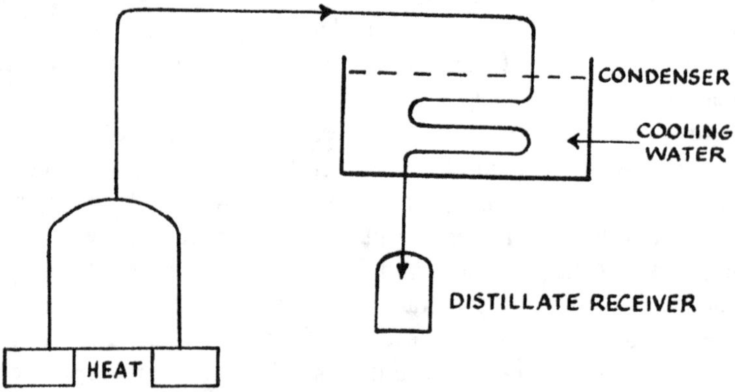

FIG. 12. Simple distillation still

Butadiene and cracking

The higher boiling fractions, generally, have larger molecules than the lower, and it was found that heating them to high temperatures caused the large molecules to break down to smaller ones. This process is called 'cracking', and since it increased the yield of the petrol-type fractions it was used for many years to augment the supply of motor spirit. Chemical examination of the cracked oil showed, however, that it often contained chemically useful compounds, and much effort has been devoted both to finding markets for these by-products and to improving the yield of them in the cracking process. Rubber chemists have always been interested in these experiments for butadiene is one of the many by-products of cracking.

Since the end of the last war there has been an abrupt shift of emphasis. Most of the crude oil imported into this country is still cracked to obtain mainly motor fuel, but a proportion is now cracked in such a way as to obtain higher yields of butadiene and other products which provided the raw materials for plastics and detergents. In this second type of cracking, petrol and fuel oil are now the by-products.

There are many different cracking processes in use in this country and in different parts of the world. For example, at the Esso Refinery at Fawley, Southampton, which supplies butadiene to the International Synthetic Rubber Company's large plant at Hythe, there are two cracking processes in use for the production of butadiene. One is called steam cracking, the other catalytic cracking. The catalytic cracking process can produce about four times as much butadiene as the steam cracking process, and the total butadiene output is in the region of 40,000 tons a year. Both processes are highly complicated and only a brief description will be attempted here.

The steam cracking process is usually performed on some of the lower boiling fractions of crude petroleum, often referred to as light distillate or light naphtha. The oil is vaporized by being passed through a series of tubes arranged in a furnace. It then meets preheated steam and the cracking process takes place in more tubes deeper in the furnace known as the 'soaker line'. The hot exit gases are partially cooled by an injection of cold oil, and the heat removed used to generate steam for further cracking. The partly cooled products then pass through a series of separating columns where tar, crude petrol, water and gaseous products are removed. The gaseous products receive further purification, and the crude petrol—itself containing dissolved gases—is passed to another separating column. Here the dissolved gases are removed from the petrol, and the petrol proceeds for further refining, eventually to become motor spirit. The dissolved gases, one of which is butadiene, go on to a special recovery unit where the butadiene is purified.

The catalytic cracking of oil is carried out by allowing the hot vaporized oil to come into contact with a catalyst, which is usually a mixture of metal oxides. The catalyst helps to speed up the cracking process, but takes no direct part in it, and can be used over and over again.

The vapour stream coming from the catalytic cracking unit contains hardly any butadiene at all, but it does contain certain near relatives of butadiene called butylenes. The butylenes have molecules similar to those of butadiene but they contain more hydrogen atoms, and one way of converting them to butadiene is to remove hydrogen from them—the dehydrogenation process. After several stages of purification resulting in a gas stream from the cracker which is mostly butylene, the dehydrogenation process is effected and butadiene is the product.

This process is also conducted with the aid of a catalyst, though a different one from that used for the cracking. The hot butylene gas is mixed with superheated steam and passed through the catalyst in what is known as a fluid-bed. The gas stream from this process is still rather dilute in butadiene and, after cooling and compressing, it receives further treatment to concentrate the butadiene, before it finally passes to the butadiene extraction unit. In the extraction unit the stream from the steam cracker and the catalytic cracker are purified side by side, the process being very similar in each case.

Butadiene intended for polymerization must be exceedingly pure: not less than 98·5 per cent. The final product of the processes just described is butadiene mixed with other hydrocarbons which have very similar properties. This makes it very difficult to purify.

One of the most successful ways of purifying butadiene found to date is to use solutions of certain copper salts. Butadiene and most of the other products of the cracking process form unstable compounds with these copper salts, and if the temperature is chosen correctly only the butadiene compound will be stable and thus left behind, while the others decompose and the gases evaporate. The solution can then be heated. The butadiene compound itself breaks down, and butadiene bubbles out of the solution, very pure. The solvent can be kept for more solution work.

The economy of recycling any component which can be used again is practised diligently in all types of chemical production. Any reaction which gives out heat is made—if required—to heat up incoming reactants by heat exchangers, or to give up heat to reactions which need it. In this way there is very little waste and the cost of the product is kept down. In the production of butadiene there are further sources of economy. The hydrocarbon gas ethylene is also produced in the steam cracking operation, and this is the raw material of one of the most important plastics—polyethylene. Motor spirit is also made. Since there are ready markets for these materials the net cost of producing butadiene is considerably lowered and, in fact, it is one of the cheapest monomers available today.

STYRENE

The other monomer of general purpose synthetic rubber is styrene. Unlike butadiene, which is a gas, styrene is a liquid. It has been known for over 100 years and its polymers for nearly as long. It occurs in both petroleum and in coal tar, the tar obtained after the dry distillation of coal in the gas works, but neither of these two sources has been found to give a sufficiently pure product for making synthetic rubber.

Today the hydrocarbon styrene is made by reacting together two other hydrocarbons, ethylene, which we have just met in connection with plastics, and benzene, which under the more familiar name of benzol is an important constituent of certain brands of petrol. The reaction product is a compound known as ethyl benzene, and this then undergoes a similar type of reaction to the one in which butylenes are converted to butadiene (in other words, a dehydrogenation reaction) to give styrene.

A very simplified diagram of the manufacture of ethyl benzene is given in Fig 13. Pure liquid benzene first enters the make-up tank A. Here it makes contact with a solid catalyst which is held in a wire basket. The benzene stream carries some of the catalyst with it and is pumped through a heater to the top of the reaction tower B. It then flows down the tower meeting ethylene gas on the way up. Reaction takes place in the tower and unreacted ethylene passes out through the top and is recycled.

The reaction product, ethyl benzene, passes down the tower and out at the bottom to a sump C. From here unreacted liquor is recycled, and the crude ethyl benzene passes to a separator D. It is then fed to the bottom of a distillation column E, and here it is distilled. Unreacted benzene passes out at the top to be recycled, and higher boiling residues remain at the bottom and are rejected.

The chemical formula of ethyl benzene is C_8H_{10}, and that of styrene C_8H_8. Ethyl benzene has thus two more hydrogen atoms than has styrene. The removal of these two hydrogen atoms, making styrene from ethyl benzene, is carried out once again by means of a catalyst. Ethyl benzene is mixed with steam and passed over the catalyst at a high temperature. The reaction takes place in tubes of

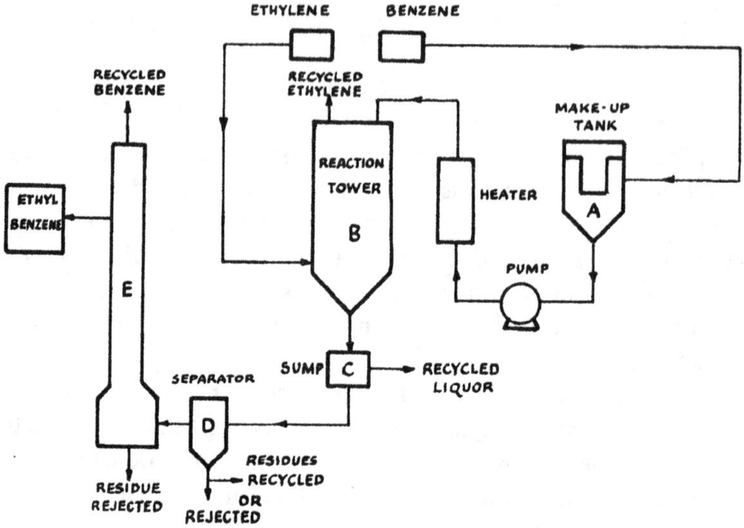

FIG. 13. Manufacture of Ethyl Benzene (Dow process)

nickel chrome steel, lined with a special alloy to prevent thermal cracking of the styrene.

The crude styrene formed in the reactors is led away and distilled to purify it. The process is difficult as the heated styrene tends to polymerize and form a solid which blocks the still. This can be overcome to a large extent by adding a small quantity of a chemical which inhibits this polymerization, known as an inhibitor or antioxidant. We shall hear more of antioxidants later. After distillation styrene has a purity of well over 99 per cent.

PRODUCTION OF THE POLYMER

The mechanism of the polymerization reaction is dealt with elsewhere in this book. All that will be said on the subject here is that provided the proper ingredients are mixed together, polymerization will proceed. The small units from the monomers will join up to form long chains and synthetic rubber will result. But to understand how the reaction is carried out at the synthetic rubber plant it will perhaps be advisable to look a little more closely at the system of polymerization generally in use today—emulsion polymerization.

Emulsion polymerization

Styrene is a clear, strong-smelling liquid, and butadiene is a gas, but butadiene liquefies readily under pressure and becomes a liquid like styrene, with a rather unpleasant smell. Neither of them will mix with water. If stirred together the mixture quickly separates out into two clear layers, one of butadiene and styrene, the other of water. But add a little soap and again stir and the result will be homogeneous and will look like thin milk. In fact, it is an emulsion. The butadiene and styrene are now in very small droplets, much too small to see, and each droplet is covered with a layer of soap molecules which keep it suspended in the water.

To start the polymerization reaction a substance called the initiator is added, and the reaction materials could be heated. Normally heating speeds up the reaction, and this is desirable as, other things being equal, the quicker the reaction the cheaper the product. But heating also has another effect. It leads to short and heavily branched chains. Chain branching is a subject which is dealt with elsewhere; here it is only necessary to note that both these effects are serious disadvantages, because, within limits, long and unbranched chains give a much better rubber. This means that the initiator must be relied on to produce a reasonable reaction rate at a low temperature. Most initiators in use today are so efficient that a good rubber can be produced at a temperature near the freezing point of water.

The polymerization formula

The list of substances used in polymerization is sometimes referred to as the polymerization formula, and sometimes merely as the 'recipe'. This is not such a peculiar choice of a word as it seems, for a polymerization formula has often much in common with a cooking recipe, notably that the choice of ingredients is based more on experience than scientific selection. No attempt will be made to give a sample recipe, as the accepted ingredients vary widely; and the following is a list of the types of components used, together with notes on their functions.

(1) *The basic phase.* The substance in which the emulsion is made is usually water, and this is called the basic phase. If the temperature at which the polymerization is to be carried out is very low some antifreeze may have to be added. Water usually comprises 60 to 80 per cent of the total system.

(2) *The first or main monomer.* This is butadiene, which comprises about 75 per cent of the total weight of monomers used.

(3) *The second monomer.* Styrene is the second monomer and makes up the remaining 25 per cent of the monomer concentration.

(4) *The emulsifying agent.* This is usually some sort of soap. Sometimes the soap may itself be made in the emulsion by adding a fatty acid and caustic soda. These react together to form soap.

(5) *The initiator.* The initiator may be a single substance or several, and may be soluble in either the monomers or in the water. The concentration may be anywhere between 0·1 and 1 per cent of the total monomers.

(6) *The modifier.* The initiator starts the reaction off. Another substance is then added which helps to ensure that the reaction proceeds correctly, that it gives chains of the desired length. This is called a modifier. The modifier also often assists the catalyst in speeding up the reaction.

(7) *The short or end stop.* The last component added is the one which stops the reaction. This might seem an unnecessary refinement, but if the reaction were allowed to go on until it stopped of its own accord, that is, until the supply of reacting materials ran out, there would be two unpleasant consequences. First of all, in spite of the modifier there would be far too many branched chains, and secondly the end would be a long time coming, for the reaction slows down as it nears the end. Thus a poorer quality rubber and—because time is money—a dearer rubber would result. Experience shows that the best results are obtained by allowing the polymerization to go about two-thirds of the way to completion.

A substance which acts as an end stopper can sometimes, as we saw in the case of styrene, be used to prevent or inhibit polymerization in the pure monomer. Inhibitors are always added both to styrene and butadiene soon after they are made, in order to preserve them in transit to the synthetic rubber plants, and on storage there.

The synthetic rubber plant

This description of a synthetic rubber plant is not meant to apply absolutely to any one plant in particular. The design and construction varies from country to country and improvements are always being made. The description is therefore meant to be typical, and to give the principles if not the details of an average plant. Most plants may be split up into five parts or areas as shown in Fig. 14, and each will be dealt with separately.

The tank farm. The monomers are stored at the tank farm. Butadiene is stored under pressure, usually in large spherical containers known as Horton Spheres with capacities of up to 250,000 gallons each. Styrene is stored in long cylindrical tanks, and since it is already a liquid it does not need to be kept under pressure.

Both monomers are highly inflammable, and mixtures of butadiene and air are explosive. For this reason the tank farm is situated some distance away from the rest of the plant, and each tank has a

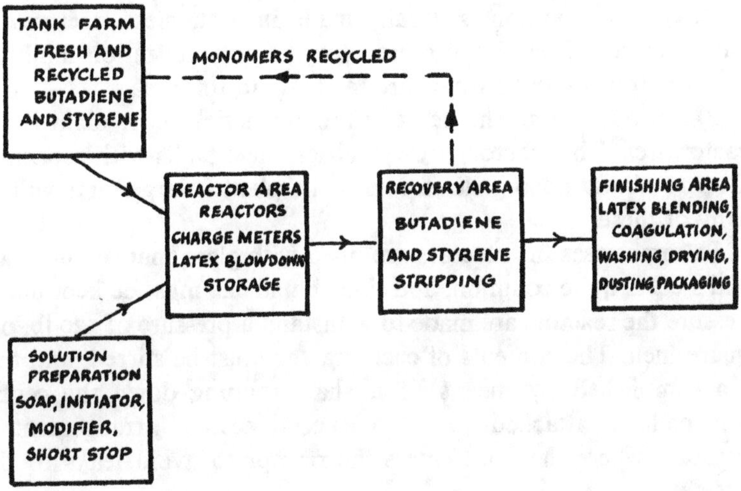

FIG. 14. Area diagram of manufacture of Butadiene-Styrene synthetic rubber

low surrounding wall so that if the monomer escapes it will not spread all over the site. In addition to this a gas detector is installed which samples the air at intervals and sounds an alarm if the concentration of monomers rises above a certain level.

There are many other precautions taken against fire at the plant. All equipment in the plant containing butadiene is provided with safety valves which discharge into a collecting system of pipes. These pipes are flushed with nitrogen, a very inert gas, which carries the butadiene along and prevents the formation of an explosive atmosphere. The pipes lead to the top of a tall chimney where the inflammable gas is burnt continuously.

The monomers are stored with inhibitors, the inhibitor being removed from butadiene before polymerization by washing with caustic soda. The inhibitor is not usually removed from styrene as its removal (which is easier) from the butadiene means that the overall concentration will be so low as not to interfere unduly with polymerization.

Solution make-up area. Here the water used in the emulsion is purified and softened, and the soap solution is made up. The soap solution is usually made in large wooden vats by adding soap chips to softened water. It is warmed before use and checked to make sure that it is neither too acid nor too alkaline.

In this area the initiator is also made up—if it is to be made up in water. The solution is usually made in a stainless steel vessel. Stainless steel vessels are also used for the polymerization inhibitor or short-stop solutions which are made up in this area.

The reactor area. The heart of the synthetic rubber plant is the reactor area. The reactors are cylindrical steel tanks with capacities of up to about 4,000 gallons, and if the plant is large there will be many of these.

In some cases the reactors are lined with glass enamel, although mild steel is quite common, and since butadiene must be kept under pressure the reactors are made to withstand a pressure of 120 lb. per square inch. The contents of each reactor must be stirred, and this is accomplished by means of a shaft running down the centre with paddles attached to it. A special sealing arrangement is necessary where the shaft enters the reactor to avoid leaks to the outside.

Heat is given out during the polymerization reaction and to maintain the emulsion at a constant temperature this heat must be removed. Reactors therefore have a jacket round the outside through which brine or cold water may be circulated, and a coil of cooling pipes on the inside. If the polymerization is to be conducted at low temperatures refrigerated brine is pumped through the coil. Careful control of the temperature of the reaction is essential, and this is done automatically.

To begin the reaction the monomers are pumped from the tank farm, and soap solution and water are pumped from the solution preparation area. The liquids pass through separate meters which are pre-set so that when the correct amount has been passed the pumps are automatically stopped. The initiator and modifier are then added and the reaction begins. Samples of product are removed from the reactor at intervals and, after drying down, subjected to tests to see if the proper physical properties have been acquired. The extent of the conversion is estimated by weighing the amount of polymer in a given volume of emulsion. When the conversion has reached 60 to 66 per cent the reaction is stopped by adding short-stop solution, and the product is ready for separation.

This system, where the reactions are run in single reactors which are emptied when the reaction has finished, is known as the batch system. Synthetic rubber can also be made by a continuous process. In this method the reactors are connected together with inlets at the bottom and outlets at the top, and the outlet of one is connected to

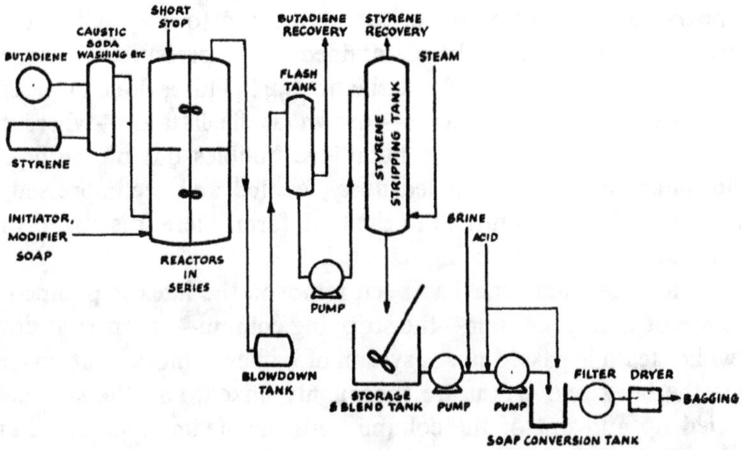

FIG. 15. Flow diagram of manufacture of Butadiene-Styrene synthetic rubber

the inlet of the next. Baffles are placed inside each reactor to ensure proper mixing and also that the liquid takes the correct path. Reactants are fed continuously into the first reactor and travel through the others while polymerization is proceeding. To increase the time of the reaction large-diameter steel pipes are sometimes connected to the last reactor in the line and polymerization can continue in there. The reaction may be speeded up slightly by increasing the temperature of the last few reactors. Further flexibility may be obtained by adding the short-stop solution anywhere in the reactor line. This method of polymerization has all the advantages associated with continuous production, and also gains from the fact that the reactants are kept continuously under pressure.

Recovery. When polymerization stops the reactors contain: synthetic rubber, unreacted styrene and butadiene, water, soap and small amounts of various other chemicals. Since the newly formed synthetic rubber is an emulsion in water—in the same condition as natural rubber in the rubber tree—it is called a latex. The latex is passed under pressure to another tank called a blow-down tank, which is sometimes glass lined, and from there it passes to the recovery area.

The polymerization has proceeded about two-thirds of the way to completion, and about one-third of the original butadiene and styrene are therefore left unreacted. These must be recovered and used again if the process is to be economic. Butadiene is fairly easily removed. At normal pressure it is a gas, and to keep it liquid for polymerization it has to be maintained under pressure. To remove it, it is only necessary to release the pressure. This is done by passing the latex through two tanks—known as flash tanks—where the pressure is released, and the butadiene bubbles out of the liquid. The butadiene gas is then led away, cooled and recompressed to make it liquid, and returned to the tank farm. Here it is mixed with fresh butadiene before being recycled.

After the butadiene has been removed the latex is pumped to the top of a large column—the stripping column—and passed down it while steam is passed up. A system of baffles in the column ensures that the latex and steam are thoroughly mixed, and the styrene is carried up and out of the column with the steam. The steam and styrene then pass to a separating vessel where the styrene separates out as a liquid on top of the water, since it is lighter. The styrene is then removed and returned to the tank farm.

Coagulation and drying. The latex, which is now referred to as stripped latex, is passed to a large vat, either of wood or concrete, where it is blended with other latices to ensure a uniform product. Next, common salt is added. This effects a partial coagulation of the latex, known as creaming because a thick cream rises to the surface. Acid is added and this coagulates the cream into large masses which look like white breadcrumbs, and also breaks down the residual soap in the latex. The crumb is then washed and filtered.

All this is carried out as a continuous operation. The creaming may take place in a pipe in which latex is passing, and a few feet farther on the acid is added. The coagulum then reaches a pump in which it is violently agitated and the result is a highly porous

Photograph: courtesy North British Rubber Co. Ltd

Tube extrusion

Latex being extruded from glass nozzles into the coagulating tank, in the manufacture of rubber thread

'Green' tyre being inserted into the mould, over the diaphragm

MOULDING A TYRE

Finished tyre being removed from the mould

crumb which is easily washed with water and does not hold back any coagulating acid in the rubber.

The crumb is washed, and filtered by being passed across a vibrating screen, which as its name suggests is rather like a flat vibrating fireguard. After washing, the crumb is dried in an oven through which it passes on a conveyor belt. Then come the automatic scales where batches of about 70 lb. are weighed. The batch is compressed in a baling machine, dusted with powdered French chalk and made up into bags. The bags are stitched automatically.

In many ways a synthetic rubber plant is a good example of an automatic factory. A small number of engineers and chemists are required to set the controls for the pumps, etc., and to check constantly the quality of the product and its ingredients. But apart from this, the process largely runs itself, and the men who cart the bags away at the end are the first to touch the product.

5

The processing of general purpose synthetic rubber

THE main general purpose synthetic rubber described in this book is a co-polymer of butadiene and styrene, made from about 75 parts of butadiene to 25 parts of styrene. Since the name 'general purpose synthetic rubber' is a rather long one we might shorten it to 'styrene-butadiene rubber', or even use the initials S.B.R. This is actually following a code, now coming into use, by which the name of a synthetic rubber can be conveniently reduced to two or three letters.

S.B.R. is transported in bales of about 70 lb. each to the factory where it is processed—that is compounded and fashioned into the finished article. Fig. 16 is a simplified diagram showing the treatment that a piece of rubber, either natural or synthetic, undergoes before it reaches the stage where articles can be made. It is obviously important that one should be able to handle a general purpose synthetic rubber with the same machinery and by the same processes as are usual for natural rubber. Otherwise a changeover from natural to synthetic rubber would involve scrapping or leaving idle a lot of expensive machinery. S.B.R. has therefore been designed to process in a way similar to natural rubber.

A question that might be asked at this stage is: why do natural rubber and S.B.R. need all this processing? The answer has been implied by much that has been written in the earlier chapters. But it might be helpful to summarize it. Raw natural rubber is not good enough for continued and reliable use, and quite unsuitable for many of the uses to which treated rubber may be put. Raw natural rubber, either crêpe or smoked sheet, is not very strong. It will not stand much heat without losing its shape, and it is adversely affected by many solvents, particularly petrol-type solvents. This is not to say that raw natural rubber is useless; Hancock and Goodyear and their

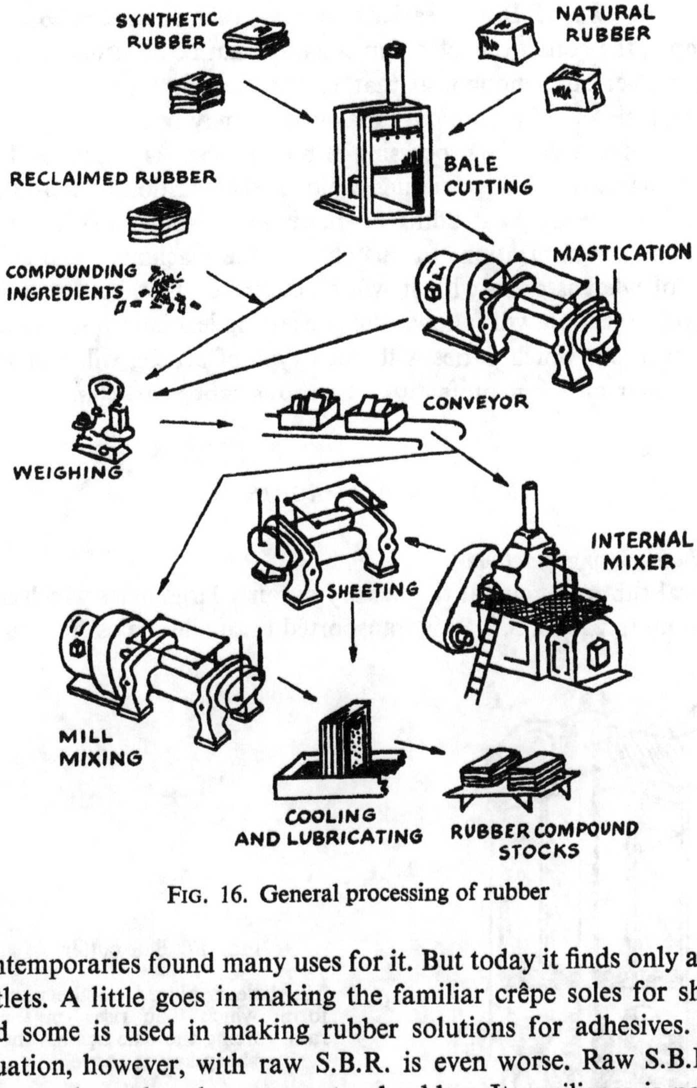

FIG. 16. General processing of rubber

contemporaries found many uses for it. But today it finds only a few outlets. A little goes in making the familiar crêpe soles for shoes, and some is used in making rubber solutions for adhesives. The situation, however, with raw S.B.R. is even worse. Raw S.B.R. is very much weaker than raw natural rubber. Its resilience is poorer and its resistance to heat is much lower than that of natural rubber.

By adding materials to natural rubber and suitably vulcanizing it, the strength and elasticity can be much improved and products made from it will last much longer. The same processes applied to S.B.R. have even more gratifying results, for by suitably compounding and vulcanizing S.B.R. the final product is very nearly the equal of natural rubber in all properties, and in some its superior.

In some ways S.B.R. is easier to process than natural rubber. For example, the conditions of polymerization can be adjusted to give a final rubber soft enough so that under the normal conditions of mixing little or no preliminary mastication may be necessary.

Apart from this the processing of both rubbers is similar and can be summarized as compounding, which is the addition of substances to rubber; sheeting, extruding or moulding or sometimes all three; and finally vulcanization. In this chapter the machinery and something of what is done with it will be described first, compounding and vulcanization will follow. The processing leading up to the final vulcanization, which varies with each type of article, will be left to the chapter on the manufacture of various rubber articles.

PROCESSING MACHINERY

The bale-cutting machine

Natural rubber arrives at the factory in fairly large bales which may weigh up to 240 lb. S.B.R. is transported in smaller bales. The bales,

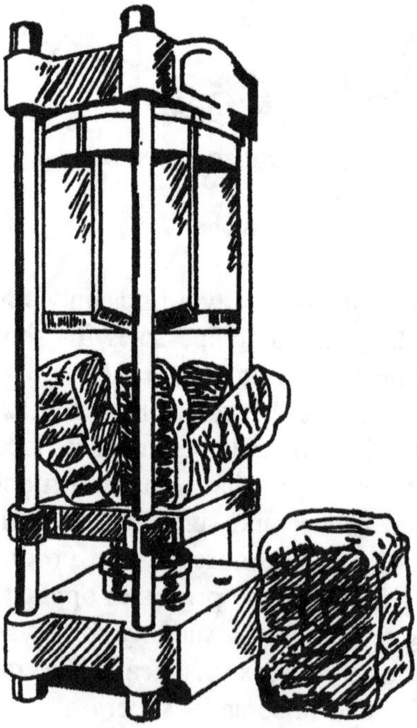

FIG. 17. Bale cutter

The bale is placed on the platform, which then rises on the ram, forcing the bale against the sharp knives above

however, are usually cut to a still smaller size by machine. The bale-cutting machine is often a kind of reversed guillotine, with very sharp blades, operated hydraulically. The bale is sliced into four to ten pieces, and with natural rubber it is usual to blend cut pieces from various bales to minimize variation in the rubber.

The mill

The mill is probably the oldest type of machine used in the rubber industry and derives from the original mixing mill of Chaffee. Modern machines are merely refinements of his mill which consisted of two cylinders or rolls placed one behind the other. Today, rolls are usually cast steel cylinders five or six feet long and two to three feet in diameter. They are hollow, and cold water or steam can be passed through them. The rolls turn towards each other and the back one revolves faster than the front. Rubber placed between them is therefore squeezed and torn at the same time, and after re-passing between the rolls several times becomes plastic and can be made to form a band round the front roll.

The space between the rolls where the rubber is nipped is called the 'nip' and this distance can be altered. By adjusting the nip the rubber can be made to form a thin band of desired thickness round the front roll, and the excess forms a 'bank' above the nip.

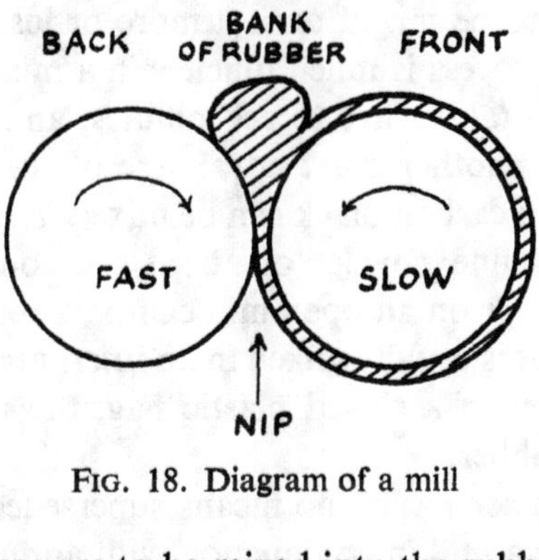

FIG. 18. Diagram of a mill

When substances are to be mixed into the rubber they are added at the bank. The rubber is then cut and peeled off as it comes upwards from the bottom of the front roll and folded back over itself as it re-enters the nip. This ensures uniform mixing.

The mill is a sort of general purpose machine in the rubber industry. It can be used, for example, to masticate rubber. Natural rubber requires from 10 to 30 minutes' treatment on the mill before it becomes plastic enough to mix substances into it, and for this purpose the back roll revolves one and a half times as fast as the front. For mixing substances into rubber the back roll does not revolve so quickly, and for making a plain sheet of rubber the rolls revolve at practically the same speed. The mill tends to be used more in small factories. In large factories much preliminary mixing is carried out in an internal mixer.

The internal mixer

A diagram of a typical mixer is shown in Fig. 19. Essentially, this machine consists of two cylindrical interconnecting chambers inside which revolve two rotors. The rotors are pear shared in cross-section and roughly spiral along their lengths. They revolve towards each other as in a two-roll mill. When they rotate the rubber is kneaded between them and between the walls of the chambers at the same time. The effective nip, therefore, travels round with the pointed ends of the rotors. To prevent rubber being squeezed up between the rotors an air-operated ram closes the gap between the chambers. The rotors and the chamber walls are hollow to allow either steam or cold water to be passed through them.

Mixing is effected merely by pouring the material down the feeding hopper, and allowing the rotors to churn up the rubber and the material to be mixed until a more or less uniform mixture is obtained. The process is much quicker in a mixer than on a mill, a complete mixing taking five to six minutes, and savings in power can be obtained. Another great advantage of the internal mixer is the ease with which carbon black can be mixed. As we shall see later, carbon black is a fine powder of about the consistency of flour. Mixing it into rubber on an open mill can be a very dirty and dusty operation. Today it is usually mixed in an internal mixer, often being added to the mixer in a closed plastic bag, the material of which disperses in the rubber.

The internal mixer has by no means superseded the mill; many manufacturers use a mill in conjunction with an internal mixer, and to improve the disperson of the compounding ingredients a final mixing is given on the mill. In any case, since the inside of an internal mixer gets very hot, sulphur is nearly always added afterwards, on the mill, to avoid premature vulcanization.

We have now reached the stage, at least in terms of machinery,

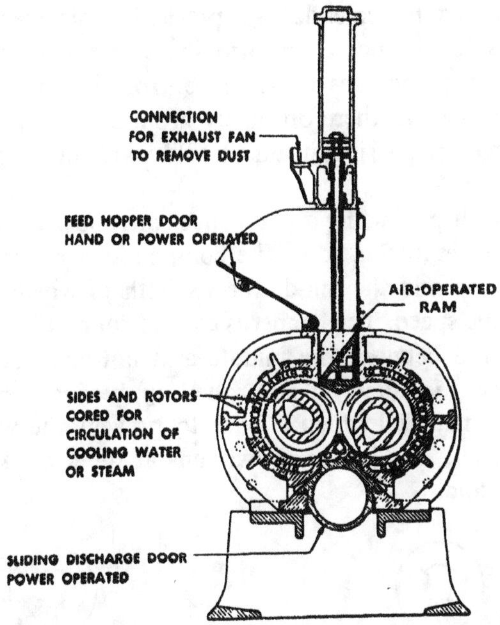

FIG. 19. Section through a Bridge-Banbury internal mixer

where the rubber has been masticated and mixed, and is now in the so-called compounded stage. No further chemical alteration of the rubber mix takes place until it is finally vulcanized. But the rubber compound may be manipulated into many shapes and sizes before it reaches a condition where it can be vulcanized. The machinery now used to process the rubber will therefore vary with the type of article which is required. But three more machines will now be described, the calender, the spreader and the extruder, because all three are of wide application and because of one them—the calender—follows directly from the mill.

The calender

The calender is in many ways the big brother of the mill. Invented by Chaffee, it is in effect a mill turned on its side so that the rolls, usually called 'bowls' in a calender, lie in a vertical plane. There are usually three and sometimes more, and each, which is often bigger than a mill roll, may weigh up to 18 tons. A further difference is that whereas the mill rolls often travel at different speeds, at least two of the calender bowls always turn at the same speed.

For all its size the calender is a precision instrument. Since one of its uses is to produce a uniform sheet of rubber of accurately known thickness, the nips must be controlled to within very fine limits. The bowls are therefore finely ground and highly polished. They are cored for passing steam or cold water in the same way as mill rolls.

A method of producing a plain sheet of rubber on a three-bowl calender is shown in Fig. 20A. The compounded rubber is inserted between the top and the middle bowls both of which are hot and run at the same speed. The rubber is carried round the middle bowl, through the nip between the middle and bottom, and round the bottom, where it is taken off and rolled up, having first been interleaved with a material called a liner to prevent the warm rubber sticking to itself. The bottom bowl runs at the same speed as the others but is cold.

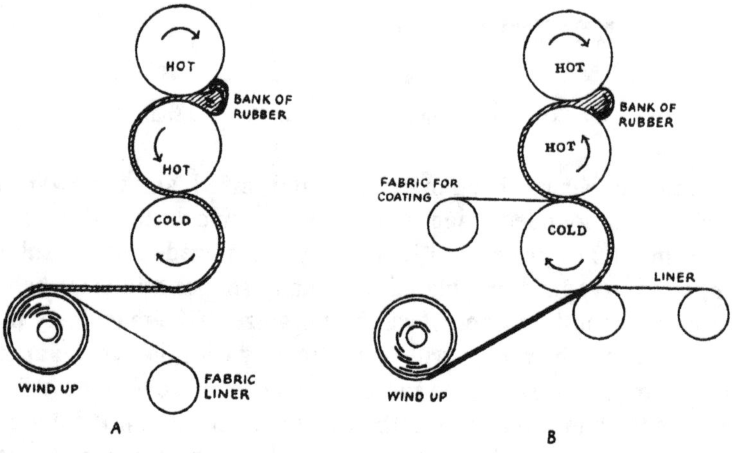

FIG. 20. Uses of a three-bowl calender

Strange as it may seem, rubber going through the nip between 18-ton bowls will actually bend them. Very little it is true, but enough to produce an uneven sheet; thicker in the middle than at the edges, for example. To compensate for this the surfaces of the cylinders are carefully ground to a suitable camber, or crown as it is called.

There are a number of other processes for which a calender can be used. One is the coating of fabric with rubber (see Fig. 20B). The rubber coming round the middle bowl is pressed on to the surface of fabric coming through from behind. This is known as top coating, or topping, and the bowls are run at the same speeds. If the middle bowl is run faster than the lower bowl the rubber may be pressed between the threads of the material, and this is known as frictioning.

The embossing of rubber sheets can also be carried out with a calender. To do this one of the bowls is covered with an engraved or embossed pattern and this is pressed on to the rubber as it goes through.

The spreading machine

The spreading machine is in some respects an alternative to the calender. There are other factors which come into it, however. The calender is often a large and expensive machine, and it takes time and rubber to adjust it, but having been adjusted it gives a high and uniform output. The spreader, on the other hand, a cheaper machine, is slower but takes smaller pieces.

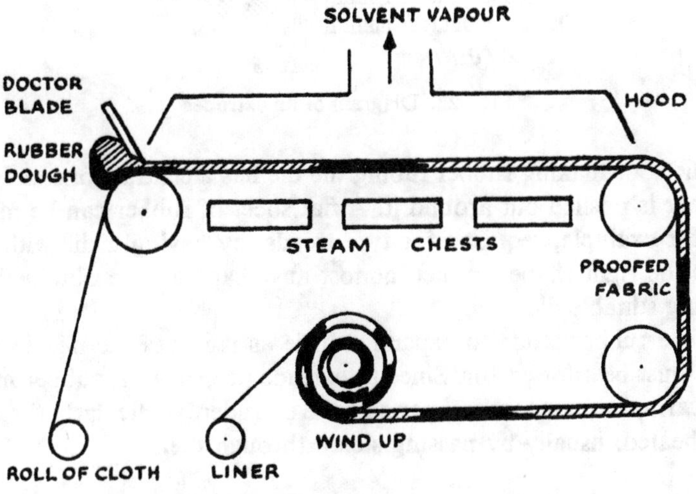

FIG. 21. Diagram of a spreading machine

For use on a spreading machine a rubber mix is made which is then soaked in a solvent and churned up in a machine called a dough mixer. This reduces the mix to a thin dough or cement. The dough is spread by placing it on a moving roll of material which is passing underneath a long knife-blade. This is called a doctor blade and it is set at an angle, as shown in Fig. 21. This spreads a film of the desired thickness on the cloth. A sandwich may be obtained by laying another sheet of material on top of the rubber, and this type of material is used to make tennis shoes.

The extruder

The last machine which we shall consider for general rubber processing is the extruder. This is somewhat like a large domestic mincing machine. A strip of rubber, usually from a mill, is fed into the hopper and is taken by the threads of the powerful screw to the end of the barrel where it is forced out through a perforated plate

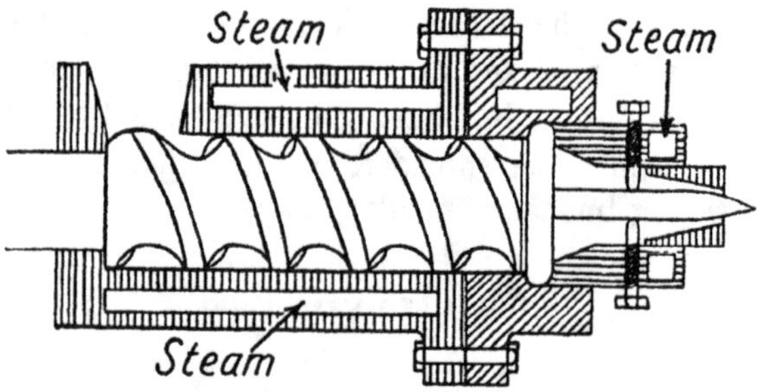

FIG. 22. Diagram of an extruder

or die. For making rubber tubing the die has a central cone and the rubber is pushed out around it. A flat sheet of rubber can be made as, for example, required for tyre treads, by having a die with the hole the right shape. In fact, almost any shape can be obtained by having suitable dies.

The rubber tends to expand a little as it leaves the nozzle and this must be allowed for. Since to extrude properly the rubber must be warm, but not too hot, parts of the extruder, particularly the die, are heated, usually by passing steam through them.

COMPOUNDING

We have touched on compounding in a number of previous chapters. To extend this, compounding might be said to be the incorporation of various substances into rubber for one or more of the following purposes: to improve or modify the properties of the rubber, to make processing easier and to reduce the cost of the article.

There are no major differences in compounding between natural rubber and S.B.R. Some slight differences exist and these will be pointed out as we go along. As a general rule anything which improves the properties of natural rubber also improves the properties of S.B.R.

The list of chemicals which can be incorporated into rubber is very large, and it has been said that almost every known material has at one time or another been mixed into rubber to see if it improves the properties. The classification of compounding substances commonly used in the rubber industry is therefore not easy, but a convenient subdivision can be made on the basis of the effect the material has on rubber. The subdivision is not very clear-cut though, and a substance may be placed in more than one category.

Plasticizers and softeners

Plasticizers and softeners are really the province of natural rubber, where they are used to help in mastication. But as different degrees of softness are often required in both natural rubber and S.B.R., depending on what further mechanical processing is to take place, softeners are sometimes incorporated into S.B.R. In practice a great many different types of materials can be used. These include vegetable and mineral oils, waxes, pitches and tars and certain resins.

A popular softener which is especially useful when natural rubber mixes for tyres are being made is pine or Stockholm tar. This is obtained from the distillation of pine wood, and is a dark brown liquid with a characteristic 'antiseptic' smell. It is also used for treating sacks and ropes. The liquid is a complex mixture of substances and contains components which protect the rubber from attack by oxygen. Apart from this and its softening properties, which are particularly useful when carbon black is being mixed into rubber, pine tar is also an excellent producer of tack.

Tack

Tack, or the property of rubber sticking to itself, was noted by Hancock, and he tried to make use of it to stick his small pieces of waste rubber together to make larger and more convenient sheets. Today tack is still useful for sticking pieces of rubber together, and it becomes very important when several pieces of rubber have to be built up into an article, as for example in the fabrication of a tyre.

The early butadiene-styrene rubbers were so poor in tack that it was extremely difficult to build tyres with them. But various modifications in the polymerization system, and the inclusion of softening agents such as certain synthetic resins which impart tack to the final rubber stock, have improved matters considerably.

Extenders

The class of plasticizers and softeners, which is already a large one, shades into another division of rubber compounding ingredients, that of extenders.

Extenders have been defined as those materials which possess plastic or rubber-like properties, and can be used to replace rubber or 'extend' the rubber mix. This definition is useful though not really wide enough, as perhaps the largest class of extenders for S.B.R., certain oils, have no elastic properties and can hardly be considered plastic. In many cases softeners may become extenders if enough of them are used, and certain inert fillers may also be classed as extenders.

The main criterion of an extender is one of cost. If a material can be mixed into rubber without affecting its properties too much, and it happens to be cheaper than rubber, then it may legitimately be called an extender.

Probably the most important extenders for S.B.R. are certain rather heavy oils. We have seen that the polymerization of S.B.R. is adjusted so that the product is fairly easily processed on normal rubber machinery. But the conditions of polymerization can be altered slightly to produce a rubber which is much stiffer and harder to process. The difficulty of processing it can be overcome by adding from 25 to 40 per cent of these oils to the rubber. This produces an easier processing rubber with very little loss of properties.

Once again the cost factor is important. The oil is relatively cheap and the rubber effectively extended. A synthetic rubber plant having an annual output of, say, 50,000 tons of S.B.R. could increase this to 70,000 tons by extending the rubber with 40 per cent of oil. The oil is not usually mixed into the rubber on the mill or in the mixer. It is made into an emulsion in water and added to the S.B.R. latex. When the latex is coagulated the rubber contains the requisite amount of oil.

Reclaimed rubber

Before we leave the subject of extenders, or plasticizers for that matter, we must mention reclaimed rubber, for this has some of the characteristics of both. Reclaimed rubber is scrap vulcanized rubber which has been treated so that it can be re-milled and incorporated into a rubber mix.

Reclaiming was begun within a very few years of the discovery of vulcanization and is today a thriving industry. It represents about 20 per cent of all the rubber consumed in America, and in Britain about 10 per cent. Since most rubber goes into tyres, reclaimed rubber is mostly tyre rubber. There are many processes of reclaiming in operation today, but they nearly all begin in the same way. First the bead is cut from the tyre, and if the tyre is thick the tread is split off. The rubber is then cut into pieces and ground on a special mill called a 'cracker'. Any ferrous metal is removed by a magnetic separator and the finely ground material is subjected to any one of a number of reclaiming processes. Perhaps the most common is to treat the scrap with caustic soda to remove the tyre fabric, and then digest with a reclaiming oil, which more often than not is pine tar. The process is effected by heating under pressure for 10 to 12 hours, after which the reclaim is washed and dried and milled again on high-powered mills. Sometimes the rubber is strained in a machine like an extruder, and then it is sheeted out to a convenient size.

The advantages of incorporating reclaimed rubber into mixes are obvious, since it contains nearly all the ingredients of the original rubber mix and being cheaper than raw rubber it is a valuable extender. It also improves the processing characteristics of the mix, enabling easier calendering and extruding. The disadvantage of using it is that it may affect the physical properties of the final rubber, and it is only used, in common with most extenders, where the highest quality vulcanized rubber is not required.

Fillers

The fillers form one of the most important classes of materials used in the rubber industry, and the class is fairly easily characterized and defined. All the substances in it are powders, usually very fine powders, and many are pure chemicals. There are no complex oily concoctions found in this class. Nevertheless it can be split into two sections, the inert fillers and the reinforcing fillers.

Inert fillers. Inert fillers are similar to extenders. They do not improve the properties of the rubber very much, but they often lower the manufacturing cost. Sometimes they impart 'body' to the rubber, that is they increase the hardness and weight, and are therefore useful in rubber mats and tiles. Examples of this class include whiting, French chalk and clay.

Reinforcing fillers. Carbon black is undoubtedly the most important reinforcing filler, and there are nearly 100 different types and makes. Carbon black, or to give it its more general name, soot, was first made by the Chinese, probably about 1600 B.C., who prepared it by burning oil and allowing the flame to play on a porcelain surface. The black was scraped off at intervals and used, mixed with glue, to make ink. Carbon black has been an important ingredient of ink ever since.

Both Hancock and Goodyear took out patents for the use of carbon black in rubber, though neither of them seems to have fully appreciated the valuable properties which it imparts. The blacks were obtained in their time by burning oil or coal gas and cooling the flames on cold plates or channels. In 1872 a black was made in America by burning natural petroleum gas, and it was called gas black.

Later on, with the greatly increased use of motor vehicles, a search was made for substances to improve tread wear. Investigations at the Silvertown Company—officially the India Rubber Gutta Percha and Telegraph Works of Silvertown, London—showed that gas black incorporated into rubber greatly increased the tread wear. This information soon spread to America, where the gas black was made, and a great new industry was born.

The general public were at first suspicious of the new black tyres, and an extensive advertising campaign was needed to convince them of their superior quality. First of all quite small amounts of black were used. In 1918 only about 12 per cent of black was used in a tyre tread. But today, in treads, 1 lb. of carbon black is used for every 2 lb. of rubber, either natural or synthetic.

The original gas black was made by burning natural gas. Much of it is still made by this method today, the gas being piped into long sheds where small jets of burning gas impinge on cool iron channels. The black is scraped off the channels and when grit has been removed is conveyed to a central packing house. It is known as channel black. If instead of small flames large ones are used, and they burn in an enclosed space without possibility of impinging on any surface, in a furnace for example, a different type of black is obtained. This is called furnace black. Another method of making black is to decompose gas by injecting it into very hot chambers. This is called the thermal process.

The furnace blacks came into prominence in America during the war because they could be made by burning oil, generally diesel fuel. The changeover was stimulated by rises in the price of natural gas. Since oil can be transported more easily than natural gas the furnace process has been worked in other countries, notably Britain, Germany and France.

Before carbon black, the chief reinforcing filler used in tyres was zinc oxide, and tyres were considered to be good if they gave 3,000 miles of wear. Today, with the use of carbon black and improved compounding, tyre-tread lives of over 20,000 miles are not uncommon. This improvement can be obtained with natural rubber where the effect of incorporating carbon black is to increase the modulus and tensile strength. Abrasion resistance and resistance to tearing are also greatly improved though the resilience decreases somewhat. With S.B.R. the effect is even more striking. The tensile strength can be increased about ten times by incorporating carbon black, bringing it up very nearly to that of natural rubber. The abrasion resistance of S.B.R. in comparable mixes with natural rubber, and containing carbon black, is rather better than that of natural rubber at high temperatures. This shows itself in hot countries where S.B.R. tyres perform better than natural rubber tyres.

Another reinforcing filler, which has been successfully used in shoe-soling compounds, is high styrene resin. There are a number of high styrene resins in use and they are made in a similar way to S.B.R. except that the co-polymer is made from about 85 parts of styrene to 15 of butadiene. Polystyrene is, of course, a plastic, a hard glass-like material, and the co-polymer made with a high ratio of styrene is much harder than S.B.R. Mixed with rubber it gives a hard compound which is very useful for soling shoes. It has also been used in rubber flooring and making castor wheels.

Colouring materials

The use of carbon black in rubber naturally imparts a black or very dark grey colour to it, and if different colours are required carbon black cannot be used. The early tyres were light-coloured, due to the use of zinc oxide, which was then used for its reinforcing properties but is now sometimes used to impart a white colour. A much better white can, however, be obtained by using titanium dioxide.

Rubbers today are usually coloured by the use of synthetic dyestuffs. The progress that has been made in the field of synthetic dyestuffs, and which results in the astonishing range of colours seen in textiles today, has also helped the rubber compounder. A very wide range of bright, fast colours is now available for all kinds of rubbers.

The ageing of rubber

One of the most important problems a rubber compounder has to face is the perishing—usually referred to as the ageing—of rubber. The catastrophic effect of ageing on raw rubber was one of the reasons which caused Goodyear and his contemporaries to search for a way of curing the rubber, and which resulted in vulcanization. Vulcanization, though it improved the properties, did not remove the cause of ageing, and in many cases it was only a temporary expedient.

Ageing in rubber is usually due to the action of oxygen. While oxygen is essential for most types of life, it is slow death to rubber; and since it is a constituent of the air, rubber is always likely to be attacked. The mechanism of the process is not fully understood, but oxygen attack undoubtedly leads to the breaking of the molecular chains, and as we saw earlier it is these long chains which give rubber its elasticity. Only very small amounts of oxygen are needed to perish rubber, since only one atom of oxygen is required for every break in the chain. If the break happens to occur in the middle, reducing the chain length by half, there will be a considerable drop in elasticity.

The attack of oxygen on rubber has, on the other hand, a number of advantages. For one thing it enables natural rubber to be masticated, for the plastic state can be achieved by breaking down the molecular chains to smaller ones in just such a way as that effected by oxygen. It is in fact extremely difficult to masticate rubber in the absence of oxygen, unless high temperatures are used.

The tackiness of freshly cut natural rubber is also attributed to the attack of oxygen, and this is a very useful effect. The ageing of natural rubber is usually accompanied by a tendency to go sticky, while S.B.R. when aged goes hard and brittle. This may explain why S.B.R. shows very little tack. Apart from this, the chemical phenomenon of oxidation appears to be similar in the two rubbers.

There are other causes of ageing in both natural rubber and S.B.R. Light, particularly ultra-violet light, and the small concentration of ozone found in the air, are both able to age rubber. The ozone attack is shown by a pattern of fine cracks on the surface of rubber, sometimes seen on the sidewalls of tyres which have been standing in the open for a long time. Some metals such as copper are insidious oxidizing agents, and therefore any dyed fabrics to be proofed with rubber should be quite free from copper salts. Of course any machinery with which rubber comes in contact while it is being processed must not contain copper.

Antioxidants

Although complete protection from all these effects cannot be supplied at present, much of the deleterious effect of oxygen can be counterbalanced by adding protective substances to rubber; and if the rubber is heavily loaded, for example with carbon black, light will not be able to penetrate very far into it, and light ageing will not be a serious problem.

The protective substances are usually called antioxidants. They were first discovered and used by German chemists working with the early types of synthetic rubber. These aged very badly and some sort of protection was most desirable. Organic bases were found to be quite efficient and developments of these are still in use today.

With S.B.R. it is particularly important to add an antioxidant early on in the manufacture of the rubber, to protect it while it is in the finely divided state, where oxygen can do the most damage. An antioxidant is therefore added to the stripped latex before the rubber is coagulated, and usually more antioxidant is added when once the rubber is compounded. Natural rubber contains some naturally occurring antioxidants when it comes from the tree in latex form. But this is not enough to protect the usual commercial grades of rubber, and further antioxidants are therefore added in a normal mix, as in S.B.R., together with the softeners and vulcanizing agents.

Vulcanization

After the addition of all the plasticizers, softeners, fillers, colouring agents, antioxidants and so on, the final treatment which converts the rubber into a desirable and useful material is vulcanization or curing. Vulcanization is often, but not always, carried out by compounding the rubber with sulphur and heating it under pressure in a mould. The heat softens the rubber and it flows until it has filled the mould. Vulcanization hardens it, and in this way an object the shape of the mould is produced. Tyres, hot-water bottles, soles and heels and many other articles are made by this method.

Natural rubber can be vulcanized by adding about 5 to 8 parts of sulphur to 100 of rubber and heating it at 140° Centigrade for from 3 to 4 hours. After this time the tensile strength of the rubber will have increased about 10 times. Continued heating now causes the strength to drop fairly rapidly and the rubber becomes brittle, then plastic again. At this stage it is obviously useless. Natural rubber vulcanized with sulphur alone, as just described, has good physical properties, but they do not last very long. Even in the presence of antioxidants the rubber will perish fairly rapidly. Luckily, however, another class of substance has been developed to help in this problem. These are accelerators, so called because they accelerate the vulcanization of rubber.

Accelerators

It appears that Goodyear and Hancock were lucky when they discovered how to vulcanize rubber, for very pure natural rubber reacts only slowly with sulphur, but in unpurified rubber the naturally occurring protein material apparently assists the vulcanization.

When experimenters began to mix substances into natural rubber they soon discovered that some of them speeded up vulcanization. Goodyear patented a few, and later lime, magnesia and litharge were widely used for the purpose. It still took a long time and much sulphur to effect vulcanization, however, and it was not until the discovery of the organic class of accelerators—lime, litharge and magnesia are known as inorganic accelerators—that the process was speeded up to any great extent.

The first organic accelerator—aniline—was discovered in 1906, but its discovery was kept secret for many years, and it was rediscovered by German chemists working on synthetic rubber. The early synthetic rubbers were much more difficult to cure than natural rubber; more sulphur was required and their vulcanized life was not very long. The discovery that certain organic bases like aniline protected the rubber from oxidation was followed by the discovery that they greatly accelerated the cure.

Since then a great many organic accelerators have been made and applied to natural as well as synthetic rubber. They are effective in very small amounts, about one part of accelerator to 100 of rubber. They have cut the vulcanization time from four hours to about 45 minutes and sometimes less, and the amount of sulphur needed from eight parts to about two or three. The smaller amount of sulphur now used and the much shorter heating times combine to give a better product, one which is stronger and lasts a good deal longer.

A whole range of accelerators is now available. There are moderately slow ones which, as described above, take about 45 minutes for a cure, some very fast ones which cure in a matter of a few minutes at quite low temperatures, and some which cure at room temperature. One or two accelerators do not even need sulphur to effect a cure, and give rise to the so-called sulphur-less cures, which have good heat and ageing resistance.

Accelerators, although improving and speeding up vulcanization, have also brought a problem which is today one of the major headaches of the rubber industry. Since vulcanization can now be brought about quickly there is a danger that it will occur too quickly. Some of the heating that rubber undergoes in processing, for example in extruding, is at temperatures quite near to those used for vulcanization, so that with sulphur and accelerators present the compounded rubber may start to cure while being processed. This is called 'scorching', and is a serious disadvantage as it may cause an article to be ruined.

In order to avoid scorching, processing temperatures are kept as low as possible, and sometimes substances are incorporated which actually retard the effect of the accelerator at processing temperatures. These are called retarders and the ideal, which is rarely, if ever, reached, is a substance which will retard at processing temperatures and not at vulcanization temperatures. Another answer is to use an accelerator which is itself ineffective at processing temperatures, but cures quickly and uniformly at vulcanization temperatures. It cannot be said that this ideal has been completely reached either.

Vulcanization activators

The prime agent of vulcanization is sulphur, small amounts of accelerators improve the cure, and there is yet another class of substances which improve the action of accelerators. They are called activators. One of the most important activators—zinc oxide—has been used for years in natural rubber, and will no doubt be used for years to come, for very few accelerators will work well without it.

Zinc oxide works equally well in natural rubber and S.B.R., the amounts incorporated being about the same in each case.

Zinc oxide is one of a class of activators known as metal oxides. Another important class is that of organic acids, the most common one being stearic acid. Here again, in natural rubber, it was a case of the activator coming before the accelerator, for natural rubber contains organic acids which activate accelerators. Nevertheless, a few parts of stearic acid are nearly always added to a natural rubber mixture.

S.B.R. also has organic acids present when it is made. These come from the breakdown of the soap used to make the emulsion, when acid is added to coagulate the latex. Stearic acid, however, is often added to the mixture before cure.

Summary of compounding

To sum up this chapter, here is an actual example of the way in which rubber compounding is worked out. This example is taken from a series of lectures given in America by the Philadelphia Rubber Group with Villanova University in 1955. The object to be made is a bumper, or door seal, for railway carriages, but it could apply equally well to the rubber seals on car doors. The bumper fits into a slot in the carriage door and the object is to cushion the door

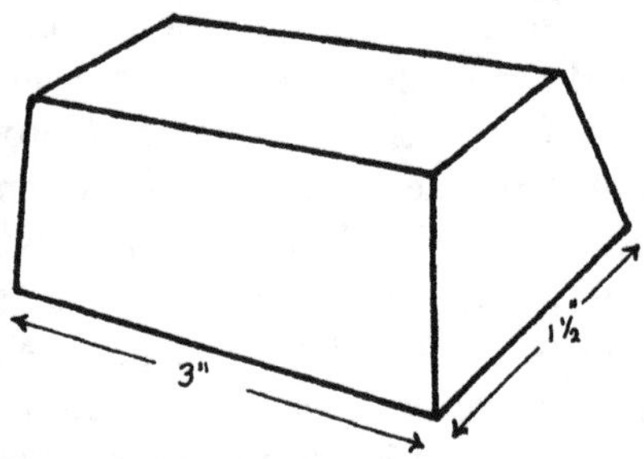

FIG. 23. Diagram of a door bumper

when it slams and keep it from rattling when it is closed. A diagram of the bumper appears in Fig. 23. It is three inches long, one and a half inches wide and wedge-shaped.

Now the bumper must be hard enough to stay in its slot when the door is slammed, and it must not 'give' too much. On the other hand it must be soft enough to cushion the door and damp out rattling when the door is closed. It must last for a few years, and it must be cheap.

The first step in deciding what compounding ingredients shall be used in the rubber mixture is to decide what the physical properties must be, in terms of things which can actually be measured; in other words the hardness, permanent set, resilience and tensile strength, etc. In deciding what hardness can be tolerated, allowance must be made for the fact that the bumper will age over the years and the hardness will alter. Then the permanent set must not be too high or the bumper will become distorted after much use. In the same way the compound must show some resilience, but it need not be very high, and this also applies to tensile strength, elongation at break and modulus, which are not quite so important and which again need not be very high.

The tolerated levels of all these properties are often decided by experience and can be checked experimentally if required. The object in this case is to produce as cheap a rubber compound as possible which is consistent with the duties it will have to perform.

Next, the manufacture of the bumper must be considered. The bumpers will be cured in moulds, and the most economic way of preparing the mixed rubber for the moulds is to extrude it, and cut it up into lengths of about three inches. The size of these 'blanks', as they are called, must be exactly known, because they must fill the mould when under pressure, yet there must not be too much overflow or rubber will be wasted. The rubber coming from the extruder should therefore be a uniform strip, and this means that the rubber should have good extruding characteristics.

A quick cure is required, for this is cheaper than a long one, yet the mixture must be such that it can be left lying about, if not required at the time, and perhaps be re-extruded later, without scorching. On the other hand, since these bumpers are rather thick, the cure should be slow enough to ensure that the outside is not cured long before the inside.

All these factors, and many others, must be taken into account when the composition of a mixture is decided upon. So it is by no means a simple matter to make up a compound which will give the cheapest and most satisfactory product.

The compounder, having weighed all the factors including temperature and time of cure, and drawing on his experience, makes up say, half a dozen trial mixes. These are cured, and the products carefully tested, before the properties are weighed against the cost and a particular mixture chosen and used.

In this example several rubbers were tried: natural rubber, S.B.R., and another synthetic rubber—butyl rubber—which we shall meet later. Clay and whiting were suggested as fillers because of their low cost, and carbon black was also tried. Oil-extended S.B.R. was also considered because of its low cost, and, having no fresh rubber there at all but using reclaimed rubber instead, was also a possibility. Listed below are four compounding formulae illustrating the principles outlined above. Each would probably do for the bumper, the final decision being made on the questions of cost, availability of the raw materials and conformation with the physical properties required.

FORMULA 1	*Parts by weight*
Natural rubber	100
Clay	100
Whiting	250
Mineral oil	25
Antioxidant	1·0
Zinc oxide	5·0
Stearic acid	3·0
Accelerators	1·2
Sulphur	4·0

FORMULA 2	*Parts by weight*
Whole tyre reclaim	200
Clay	50
Whiting	100
Antioxidant	1·0
Zinc oxide	5·0
Sulphur	3·0
Accelerator	1·0

FORMULA 3	*Parts by weight*
S.B.R.	100
Clay	100
Whiting	200
Mineral oil	60
Antioxidant	1·0
Zinc oxide	5·0
Sulphur	5·0
Accelerators	2·1

FORMULA 4	*Parts by weight*
S.B.R. (oil extended)	100
Clay	60
Whiting	150
Antioxidant	1·0
Zinc oxide	5·0
Sulphur	4·0
Accelerators	2·1

6

Special purpose rubbers

IN THE winter of 1919 an American chemist was trying to make mustard gas by a new method. He obtained instead an evil-smelling concoction with quite distinct rubber-like properties. After several years spent in trying to get rid of the smell and improve its properties, the material was abandoned. Some years later, in 1924, another American chemist, J. C. Patrick, this time trying to improve anti-freeze, stumbled across the same thing. His efforts, however, were successful, and in 1930 a new rubber appeared on the market. It was called Thiokol and was remarkably resistant to attack by oils. This was followed in 1931 by the introduction of another synthetic rubber from America called, eventually, Neoprene.

The introduction of these new synthetic rubbers, and more were to follow, caused a change of outlook among rubber manufacturers. In common with most other synthetic rubbers known at that time, Thiokol and Neoprene did not compare very favourably with natural rubber in the 'rubbery' properties. But both of them retained their properties much better when in contact with oils, and Neoprene behaved better than natural rubber when exposed to sunlight and ozone for long periods.

Here then was something new; rubbers which could be successful where natural rubber failed. Manufacturers were not slow to realize this, and though both of the synthetic rubbers were more expensive than natural rubber, and Neoprene still is, they found a ready sale. On the one hand research was directed at finding a cheap but effective substitute for natural rubber, and on the other there was a drive to find new rubbers which could be put to new uses.

In this chapter we shall look at some of these special purpose rubbers. There is not enough space to include them all, for there are a great many—some of them not yet out of the experimental stage—

and their number increases almost daily. But the main ones are not likely to change appreciably in the next few years, and we shall be concerned with these.

NITRILE RUBBER

Nitrile rubber was not the first special purpose rubber made, nor is it the most important one in terms of consumption. But it seems convenient to deal with it first because it is made in a very similar way to S.B.R. It is, in fact, chemically closer to S.B.R. than any of the other special purpose rubbers.

This rubber was first made in Germany in 1930. Since it was a co-polymer of butadiene and acrylonitrile, made with either 25 or 35 per cent of acrylonitrile, it was called Buna N. This recalls the butadiene-styrene rubber of about the same period which contained 25 per cent styrene to 75 per cent of butadiene and was called Buna S. The nitrile rubber containing 25 per cent of acrylonitrile was afterwards called Perbunan, and the grade containing 35 per cent was known as Perbunan Extra.

In Germany at this time extensive claims were made for the rubber. It was reported to be better than natural rubber for resistance to ageing, light, ozone, heat, oils and abrasion, and it also had low permeability to gases. The chief obstacle seemed to be its high cost, which was estimated to be 3*s.* 7*d.* per lb. whereas natural rubber was selling at between 6*d.* and 1*s.* per lb. With the possibility of a coming war the German Government subsidized a Buna company, and a duty was placed on the import of natural rubber.

Manufacture

Nitrile rubbers today form a separate class of special purpose rubbers. It is usual to make them from butadiene and from 18 to 45 per cent of acrylonitrile. The most frequent formulation is 30 per cent of acrylonitrile and 70 per cent butadiene. Increasing the proportion of acrylonitrile used in making the co-polymer usually increases its oil resistance. But using more than about 50 per cent of acrylonitrile produces a co-polymer which is a leathery plastic.

In many ways the preparation of nitrile rubber is very similar to that of S.B.R. Both involve emulsion polymerization in water and the soaps, initiators and modifiers are also identical or very similar. As with S.B.R., polymerization can be conducted at about 50°

Centigrade or, as is more usual now, at lower temperatures in order to obtain a rubber with better properties. To terminate the reaction a shortstop is added, and the unreacted butadiene and acrylonitrile are recovered by reducing the pressure and distilling them.

The coagulation of the latex is effected in a very similar way to that for S.B.R., by adding brine. The crumb is then filtered, washed and dried, and finally baled.

Processing and compounding

In processing and compounding there is again a similarity between nitrile rubber and S.B.R. The raw rubber has the same low tensile properties that characterize S.B.R. and in the same way carbon black greatly improves these properties. Vulcanization can be effected with sulphur and accelerators. Zinc oxide is necessary for the best results, and antioxidants are also needed.

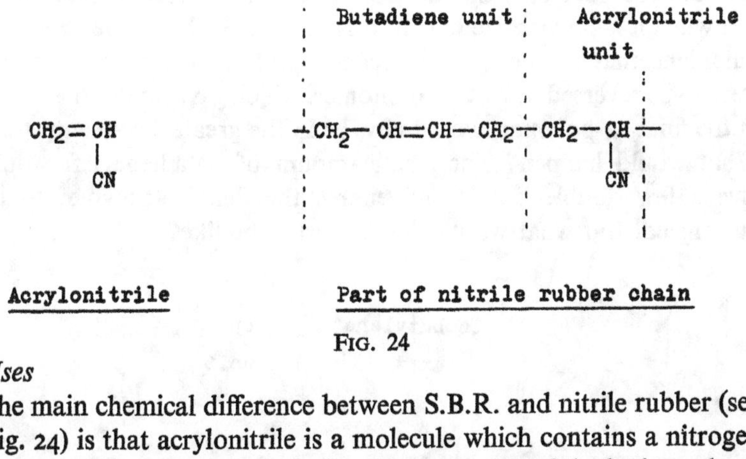

Acrylonitrile   Part of nitrile rubber chain

FIG. 24

Uses

The main chemical difference between S.B.R. and nitrile rubber (see Fig. 24) is that acrylonitrile is a molecule which contains a nitrogen atom as well as the carbon and hydrogen atoms of the hydrocarbons butadiene, isoprene and styrene. It is the nitrile group (the carbon atom and its adjacent nitrogen atom constitute the nitrile group), in the acrylonitrile part of the polymer, which gives the rubber its special properties and these are best shown by its resistance to attack by any of the crude oil fractions.

To sum up then, nitrile rubbers show their usefulness in applications where high solvent resistance is required, for example in petrol hoses and gaskets. Latices of nitrile rubber are used for impregnating paper, textiles and leather in cases where oil resistance is required. Another advantage of nitrile rubber is its compatability with other rubbers or plastics. Useful blends can be made, for example with Thiokol and high styrene co-polymers, certain plastics such as P.V.C. and the Bakelite type resins.

BUTYL RUBBER

So far all the synthetic rubbers we have discussed have contained butadiene. In fact, butadiene has been the principal monomer, the other being styrene or acrylonitrile. Now as we saw earlier, butadiene is like isoprene. They both contain two double bonds. When polymerized, one double bond per molecule disappears, leaving only one double bond for each monomer unit in the final chain. These double bonds, however, enable the polymer to be vulcanized. Styrene and acrylonitrile have only one double bond in the molecule and this is lost when it is polymerized. Thus it is only through the agency of the butadiene that the co-polymers can be vulcanized. If the position were to be reversed, the other monomer, which gave no double bonds in the final co-polymer, would now be in the greatest concentration. What would happen? The small amount of butadiene left would give a few double bonds and enable the final co-polymer to be vulcanized. But what would the properties be like?

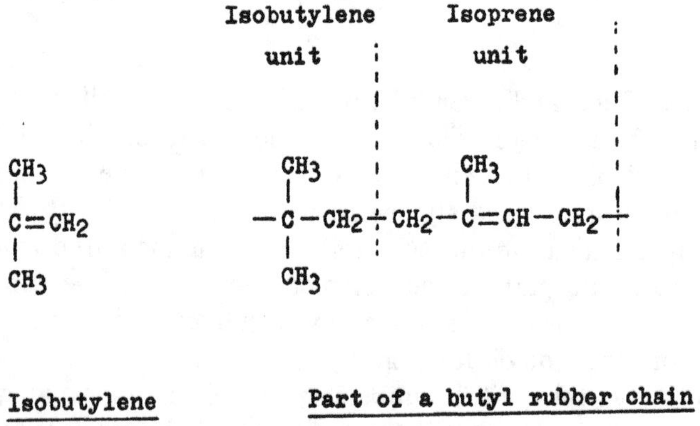

FIG. 25. (There will be many more isobutylene units than isoprene units, since butyl rubber contains only a small percentage of isoprene)

To discover this we turn to butyl rubber. This rubber contains no butadiene, but it does contain a little isoprene so the product still has a few double bonds left for vulcanization. The main monomer, however, is neither styrene nor acrylonitrile, it is a substance called isobutylene. Isobutylene is a hydrocarbon like styrene, butadiene or isoprene, but like styrene it contains only one double bond and this is lost in polymerization.

Isobutylene polymers have been known for many years. Some of them are liquids and some rubbery solids. They cannot be vulcanized, that is one cannot join the chains together by crosslinks at a few points, and therefore they tend to be rather plastic and flow slowly even at room temperature. But like most plastic materials their resistance to ageing is exceptionally good.

Adding a small percentage of isoprene to isobutylene, however, enables vulcanization to take place; crosslinks are formed and the plastic becomes a rubbery solid with very little tendency to flow while still retaining much of its original high resistance to oxygen attack.

Manufacture

Butyl rubber is made from about 97 per cent isobutylene and 3 per cent isoprene. The isobutylene is a liquid and is obtained from the cracking of oil. It is carefully purified, mixed with isoprene in the correct proportions and then with another substance, methyl chloride, which is merely to dilute the mixture. This is then cooled to —96° Centigrade, a temperature lower than that reached anywhere on earth outside laboratories. The cooled feed is led to the bottom of a reactor which is also very cold, where it meets a solution of the initiator and polymerization proceeds.

The polymerization is instantaneous and a large amount of heat is given out; in fact the reaction is almost explosive. This therefore calls for very efficient cooling. The low temperature of the reaction is necessary to obtain a good polymer with long chains.

The polymer itself is formed as very fine particles which are insoluble in the reaction medium. To remove the excess monomers the mixture is passed through an overflow line to a tank containing hot water, where the solvent and excess monomers are vaporized and led away. Final traces of the monomers are removed by applying a vacuum. The suspended butyl rubber is then filtered off, sometimes by a vibrating screen, and dried in a tunnel dryer.

Processing and compounding

Butyl rubber does not usually need to be masticated. Plasticizers alter the softness in the same way as with natural rubber, although not all that can be used for natural rubber are suitable for butyl. Reinforcing fillers do not seem to have the same effect as in S.B.R. and nitrile rubbers, and the tensile strength usually decreases as the amount of filler is increased.

A recently introduced treatment of heating butyl with carbon black in the presence of certain chemicals improves the reinforcing effect of the black, and both tensile strength and abrasion resistance are increased.

Normally butyl can be vulcanized with sulphur and accelerators, but the process is slow, owing to the small number of double bonds remaining in the rubber. Fast-curing accelerators and high temperatures are therefore used. Zinc oxide is required in all butyl compounds, but fatty acids, unlike natural rubber and S.B.R., are not necessary. The slow-curing characteristics of butyl have another important effect. Quite often synthetic rubbers are mixed or blended either together or with natural rubber, to try to improve the properties of both. But owing to the time taken to vulcanize butyl it is difficult to blend it successfully with natural rubber or any synthetics, except Neoprene. For example, if butyl rubber is vulcanized with natural rubber the sulphur put in to vulcanize the butyl will all be taken by the natural rubber before the butyl has time to use it.

Properties and uses

There are two important properties of compounded or vulcanized butyl rubber which make it different from natural rubber. Its permeability to air is only about one-tenth that of natural rubber, and because of its lack of double bonds it is chemically inert in comparison with natural rubber. It resists attack by oxygen and ozone better than Thiokol, S.B.R., Neoprene and nitrile rubbers, and much better than natural rubber.

Apart from these two factors it is much the same as natural rubber. It differs from S.B.R. in having a high tensile strength without any reinforcing fillers, but its resilience is rather poor compared with natural rubber. For these reasons butyl has some claim to be a general purpose rubber, though its more striking differences put it in the special purpose class.

Owing to its high impermeability to air and gases, butyl finds much use in inner tubes. Indeed, until the tubeless tyre removed the need for inner tubes, most of the butyl rubber used went in these. It also finds uses in electrical wire insulation, in high pressure steam hose and in some proofed fabrics.

Butyl has also been used for tyres. There is little doubt that tyres made from butyl rubber give a much quieter ride and easier braking, and of course their resistance to weathering is outstanding. Butyl rubber is, however, a rather 'dead' rubber with less resilience than either natural rubber or S.B.R., and tyres made from it require more power, at least at slower speeds, to roll them along the road. Some of the newer chemically modified grades of butyl have shown marked improvement over the older type, and butyl tyres may yet compete on more equal terms with natural rubber and S.B.R.

NEOPRENE

In 1925 a certain Father Nieuwland, professor of organic chemistry at Notre Dame University, presented a paper to the American Chemical Society in New York on the polymerization of acetylene. Not a world-shaking event, one would think, but certain chemists from the American chemical firm of du Pont who were present were very interested. Du Pont's were interested in synthetic rubber, and, in particular, making rubbers from acetylene. Father Nieuwland had produced small amounts of a derivative of acetylene called mono-vinyl-acetylene, and the chemists from du Pont's tested this for its rubber-producing qualities. The work was not very successful, but by reacting it with hydrochloric acid another chemical was produced which proved to be much better, and which polymerized to give quite promising rubbers.

This substance is called 2-chlorobutadiene and its formula is given on page 94. It can be seen that it differs from butadiene by having a chlorine atom—denoted by the letters Cl—in place of one of butadiene's hydrogen atoms.

The du Pont chemists called this substance chloroprene, and the rubber they obtained from it they at first called Duprene and later Neoprene. The rubber was first sold in 1932. Its price was 5*s*. a lb., and at that time natural rubber was selling at 1½*d*. a lb. But even then

Chloroprene units

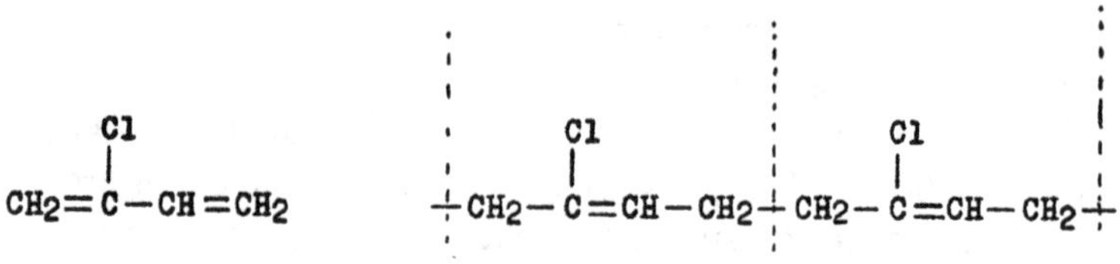

Chloroprene — Part of a Neoprene chain

FIG. 26. (Chloroprene differs from butadiene only by having a chlorine atom, Cl, in place of one of the butadiene hydrogen atoms)

it sold well—it was one of the first oil-resistant rubbers—and has continued to do so because it has certain marked advantages over natural rubber.

Manufacture

Chloroprene is emulsion polymerized in a rather similar way to S.B.R. It has a soap to emulsify the chloroprene in water, an initiator to begin the polymerization reaction, a modifier, which incidentally is usually sulphur, and a shortstop to end the reaction. Polymerization is conducted at a temperature of 40° Centigrade, a little higher than is normal for S.B.R. today, and a latex is produced. A stabilizer is added and the latex allowed to stand for a time. It is then acidified and the final coagulation is effected by cooling. A large drum filled with cold brine dips into the latex which coagulates or freezes on its surface. As the drum rotates the rubber is stripped off with a knife. The film of rubber is then washed with water, squeezed through rollers to remove most of the water, and dried by heating it in air. The dried film is gathered together to form a rope and cut into short lengths for bagging. The sticks are flexible and light amber in colour. Sometimes chloroprene is co-polymerized with isoprene.

Properties

Neoprene can be processed on conventional rubber equipment, but normally needs chemical additives which help to plasticize it during milling. High temperatures must be avoided, and in general Neoprene is not so easy to process as natural rubber. A further disadvantage is that it tends to stiffen on standing.

The main chemical difference between Neoprene and the other rubbers described is that it contains chlorine atoms. This gives Neoprene certain distinctive properties. It can be vulcanized by heating alone, although better results are obtained if vulcanizing agents are used. Sulphur can be used for vulcanization, but this is more effectively accomplished if zinc oxide is used, or a mixture of zinc oxide and magnesia. Some accelerators can be used, but the usual ones for natural rubber, S.B.R. or nitrile rubber are not effective in Neoprene and special accelerators have been developed.

When vulcanized, Neoprene compounds have a high tensile strength without reinforcing fillers, and in this respect they resemble natural rubber rather than S.B.R. Neoprene is more like natural rubber than S.B.R. in many ways. It gives latices, for example, which can be used for dipping, as the coagulated rubber is nearly as strong as natural rubber.

One of the important special purpose properties of Neoprene is its resistance to solvent attack. This is not as good as in Thiokol or nitrile rubbers, but there is compensation because Neoprene is a better rubber generally than either of the others. Neoprene has much better resistance to ozone, oxygen and sunlight than nitrile, and another very important property of Neoprene is its flame resistance. All organic compounds, that is compounds containing mainly carbon and hydrogen, burn if heated to a high enough temperature; and the rubber hydrocarbons are no exceptions, as evidenced by the number of rubber fires one hears about. For this reason, and for its other special properties, Neoprene is put to good use in hose, conveyor belts, wire insulation and mechanical goods.

THIOKOL

Thiokol rubber or Thiokols, for there are a number of these rubbers available, should really be called polysulphide rubbers. The chemical structure is of molecular chains made up of alternate groups of carbon and sulphur atoms, and an example of the repeating unit is given on page 96.

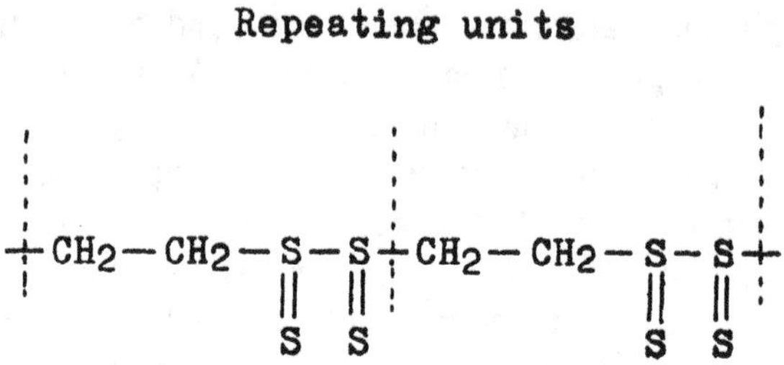

FIG. 27. Fragment of a Thiokol or polysulphide chain
The sulphur atoms are denoted by the letter S

Manufacture

The preparation of these polymers involves the joining together of pairs of molecules, one molecule to form the carbon part of the chain and the other the sulphur or sulphide part. A simple example is the chemical known as ethylene dichloride which eventually forms the carbon part, and sodium tetrasulphide which forms the sulphide part. A chemical reaction links pairs of these molecules together to form a long chain.

The ethylene dichloride is added slowly to a vigorously stirred solution of the sodium tetrasulphide in water. At the end of the reaction the polymer settles out of the suspension. It is then washed and dried and pressed into bales or slabs.

Processing and properties

There are several different types of polysulphide rubber. Most of them can be masticated on a mill in the presence of softening agents, although some of them do not need preliminary mastication. Reinforcing fillers are often incorporated into the rubber, and these, such as carbon black, markedly improve the properties, which are rather poor compared with other rubbers. The tensile strength, for example, is one of the lowest of all the synthetic rubbers. Vulcanization can be effected with zinc oxide.

Polysulphide rubber is perhaps the least rubber-like material we have met so far. Chemically it is quite different from natural rubber and in physical properties it is greatly inferior. But it has the most remarkable resistance to solvent attack of all the synthetic rubbers. A piece of vulcanized Thiokol rubber left immersed in several solvents for six months was hardly affected at all by most of them. This property has led to its use in petrol-hose, fuel lines, printers' rollers, cable-covering and diaphragms and gaskets which have to be exposed to solvents.

Another very important property of these rubbers, which is unique among synthetic rubbers, is that the long chains can be selectively broken down by certain reagents. Some of the sulphide linkages are broken and the solid polymers become liquid or putty-like without losing their essential polysulphide structure. The short chains can then be built up again by using other reagents—zinc oxide is one of them—and the liquids become solid again. This property is so useful that polysulphide rubbers are used in numerous caulking and sealing operations where solvent resistance is not so important. The unvulcanized liquid or putty-like polymer is pressed into position and provides a tight solid seal when vulcanized.

SILICONE RUBBERS

The name 'silicone' was first used over 50 years ago by an eminent English chemist, F. S. Kipping, for a compound containing the elements carbon, silicon and oxygen. He proposed the name silicone because the last part of the word—'one'—is used to describe a particular arrangement of oxygen in the molecule. Unfortunately the molecule was later shown to have a different arrangement of silicon and oxygen, but the name has stuck, and is used today to describe a range of materials containing these elements. Kipping is also sometimes quoted as an example of the danger of prophesying on scientific matters, for in 1937, as a pioneer in the field of silicon compounds and possibly the greatest authority on silicon chemistry at the time, he said that the prospects of useful commercial advances in this field were not very hopeful. Seven years later a whole range of commercially useful silicon products appeared, among them silicone rubbers.

Silicon is not a very familiar element to people who are not chemists, but it is very widespread. Most of the rocks of the earth's crust are silicates—compounds of silicon and oxygen—and include sand, clay and mica. In many ways silicon is a similar element to carbon and can form long chains of atoms, although silicon chains are usually of alternate silicon and oxygen atoms. A unit of silicone rubber, as illustrated on page 98, is a very simple one. There are no double bonds, and the rubber is therefore resistant to attack by oxygen and many other reagents.

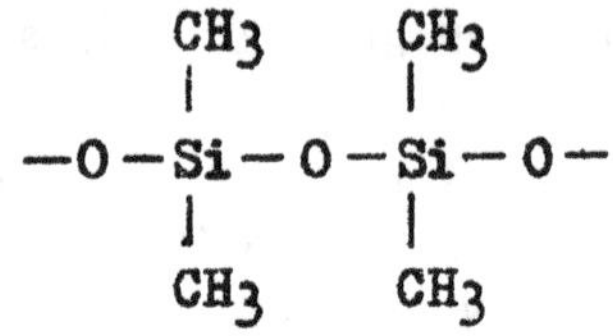

FIG. 28. Fragment of a silicone rubber chain
The main chain is made up of alternate silicon (Si) atoms, and oxygen (O) atoms

An account of the preparation of silicone rubbers will not be attempted here, as it cannot be explained without a lot of chemical details, and for the reader who requires further information a number of excellent textbooks are listed at the end of this book.

There are not very many types of silicone rubbers available today, but most are very peculiar substances. Some do not look much like rubber at all, but more like putty, for they are white and can be moulded easily with the fingers, and if placed in a container they slowly take its shape. They exhibit what is known as 'cold flow'. But if moulded into the shape of a ball and dropped, they bounce just like ordinary rubber. A type of silicone rubber possessing these unusual characteristics to a marked degree has become known as 'bouncing putty'.

The early types of silicone rubbers had very poor elastic properties. The tensile strength was low, and the resistance to tearing was poor. But improvements in compounding have increased the tensile strength until it is somewhere between that of Thiokol and nitrile rubber. Fillers can be incorporated into silicone rubbers, but carbon black is not used. A form of silica (silica is similar chemically to pure sand) can be used as a reinforcing filler, and this materially improves the properties of the rubber.

Since silicone rubber contains no double bonds it cannot be vulcanized with sulphur in the normal way. But certain other substances are capable of vulcanizing it, though the process takes several hours. Sometimes silicone rubbers are deliberately made with double bonds to enable them to be vulcanized with sulphur.

The remarkable thing about silicone rubber is that whereas its rubbery properties are not startling it retains them through a surprising range of temperature. It is still a rubber at temperatures so low as to make natural rubber and most synthetics hard and brittle, one might almost say frozen solid, and it retains these properties nearly up to the melting point of lead. Since these polymers also possess excellent resistance to steam, ozone, light and outdoor weathering, they find applications in many fields, which include electrical insulation for high temperature service, gaskets and O rings for high temperature oil, fuel and steam pipes. A particular example is their use in gaskets for steam irons, where they come into contact with the steam and form an effective seal. At the other end of the temperature scale silicone rubbers form effective elastic seals in refrigeration equipment.

POLYURETHANES

Polyurethane rubber will be the last synthetic rubber described in this chapter. It represents a triumph for pure chemistry rather than for chemical engineering. Many synthetic rubbers, in particular S.B.R. and butyl rubber, represent notable achievements in the field of chemical engineering, where the chemical reactions involved are relatively simple, although translating them into plant production has been difficult. Polyurethanes are the first real example we have met where the skill of the organic chemist has been applied in building up a rubber from simple molecules very nearly piece by piece. As we shall see later, the modern approach to the subject is towards tailor-made polymers, rubbers which are built with specific ends in view. They therefore have special properties and can do special jobs. This applies not only in the field of rubber, but in the allied fields of plastics and synthetic textiles.

There are quite a number of different polyurethane rubbers made today. The first was probably the German Vulkollan rubber made in about 1940. A polyurethane rubber was developed in England in the 1940's and there were several in America. Although these rubbers differ chemically amongst themselves they are all made in basically the same way.

The first step is to make a short chain. This, in the Vulkollan rubber, is what is called a polyester chain. There is another type in use today called a polyether chain. Both materials are usually waxy substances, and the chains contain from 10 to 30 repeating units.

The next step is to join up the ends of the chains to make longer ones. This is done by making use of a molecule with two reactive groups called isocyanate groups, each of which can join up with one end of a polyester or polyether chain, thus joining two chains together. When the isocyanate group reacts with the ends of the chain, it forms what is called a urethane group and this is what gives the rubber its name. The waxy material is usually melted and the isocyanate linking agent stirred into it, forming a syrup.

The last, or vulcanization, stage is accomplished by adding a chemical, called a glycol, which helped to make the original short chain. This is simply stirred into the syrup which is then poured into a mould, and in a very short time, often less than a minute, the syrup solidifies. After further heating the rubber is removed from the mould, fully vulcanized. It is very doubtful, however, if the vulcanization in this instance involves any chemical crosslinking of the chains.

This type of polyurethane rubber is known, not unnaturally, as cast polyurethane. There is another type, known as millable polyurethane rubber, and this is made by using not quite so much isocyanate linking agent in the second stage as for cast rubber, so that shorter chains result and the product is millable like natural rubber. To vulcanize this type, more isocyanate is merely milled in and the curing completed by heating. Alternatively, the millable polyurethanes can be made with double bonds in them, when they can be cured with sulphur in the normal way.

Properties and uses

The cast products can be made soft enough for printing rollers or harder than the hardest 'soft' natural rubber compound. Their tensile strengths are higher than that of natural rubber and they are superior in oil resistance. Fillers do not increase their strength, and have, in fact, a harmful effect. Resistance to abrasion is extremely good under some conditions, for example at normal temperatures, but falls off under severer conditions; in car tyres, for example.

Millable polyurethanes do not have such outstanding properties as the cast type, but the ability to use conventional rubber machinery is often an advantage. The properties of those cured with sulphur are improved by fillers such as carbon black.

The castable polyurethanes are very useful for making moulded goods, and since the rubber sticks easily to metal many machine parts can be made, for example, gears, valve seals, moulded diaphragms and membranes for pumps. Typewriter rollers have also been cast and a large flanged coupling for a ship's propeller has been made.

In many ways polyurethane rubber is ideal for shock absorbing, for it combines hardness with elasticity. It can be used as a head for a mallet which will drive a nail into oak yet not mark the surface of a polished desk.

Polyurethane rubber has also been used successfully for tyres on small industrial trucks where its life is much longer than that of natural rubber. In motor-car tyres, however, it cannot compete with natural rubber, S.B.R. or butyl, for its abrasion resistance falls off rapidly as the speed of the tyre increases.

Perhaps the largest use of polyurethane rubber today is in foam rubber. If to the second stage syrup, made from a polyester or polyether chain and a fairly large quantity of isocyanate, water is added, it will generate large quantities of carbon-dioxide gas which bubbles out through the syrup, and at the same time the syrup is vulcanized. The bubbles then become fixed, giving a foam rubber. The tensile strength of polyurethane foam is greater than that of foam from natural rubber latex, but it is not usually so resilient. Its permanent set is poorer but it resists oxygen attack better, and it is much lighter than natural rubber latex foam. It is difficult to mould properly, however, and is usually cast into blocks or sheets and cut to size afterwards.

HARD RUBBER

This chapter will be concluded with a section on a class of products made from rubber, both synthetic and natural: the hard rubbers.

Hancock was probably the first person to produce hard rubber, certainly he was the first to patent it. He discovered it accidentally when he left some of his strips of rubber too long in molten sulphur. The product that Hancock obtained was hard and black, and it was later called ebonite because it resembled Ebony. Another name for hard rubber is Vulcanite.

It is made by heating rubber with a lot of sulphur for long periods. The maximum amount of sulphur which rubber will take up, and this applies to natural rubber and S.B.R., is 32 per cent for natural rubber and 29 per cent for S.B.R. It appears that at these levels all

PROPERTY	NATURAL RUBBER	S.B.R.	BUTYL
Tensile strength (unvulcanized) in pounds per square inch ..	3,000	very low	2,000
Tensile strength (vulcanized) with reinforcement in pounds per square inch	800–4,500	800–4,000	800–2,500
Elasticity	high	high	fairly low
Resistance to ozone	moderate	low	high
Resistance to sunlight	low	low	high
Resistance to flame	low	low	low
Resistance to heat	fairly low	fairly low	moderate
Resistance to cold	high	high	high
Resistance to solvents	low	low	moderate
General resistance to ageing ..	moderate	moderate	high
Abrasion resistance	high	high	moderate
Tear resistance	fairly high	moderate	moderate
Resilience cold	high	moderate	low
Resilience hot	high	moderate	fairly high
Ease of processing	high	high	moderate
Tack	high	low to mod.	fairly high
Ease of vulcanization	high	high	low to mod

the double bonds in the rubber have disappeared and these represent the ultimate stages of vulcanization. Rubber in this state is hard, and more like a plastic than a rubber. Until it is finally vulcanized, however, its processing is similar to that of a normal rubber mix.

Hard rubber compounds are not specialized and their only difference from soft rubber mixes is in the amount of sulphur used, which is about 40 per cent. The quality of the raw rubber has little effect on the mechanical properties of the finished hard rubber, and for this reason low grades of natural rubber can be used. S.B.R. and nitrile rubbers give similar hard rubbers. Nitrile mixes are easier to process than S.B.R., they are less brittle when vulcanized, but have poorer electrical properties.

Sulphur is the only vulcanizing agent used. Accelerators may be used, but the time of vulcanization can only be safely cut from six or eight hours to about four hours. If the reaction proceeds too fast

THIOKOL	NITRILE	NEOPRENE	SILICONE	CAST POLYURETHANE
very low	very low	800–3,000	very low	—
600–1,500	800–2,500	800–3,000	400–1,000	4,000–6,000
low	moderate	moderate	fairly low	moderate
high	low	fairly high	high	high
moderate	moderate	high	high	high
low	low to mod.	high	high	low to mod.
low	mod. to fairly high	fairly high	very high	fairly low
ow to mod.	low to mod.	fairly high	very high	fairly high
very high	high	high	moderate	moderate
high	high	high	high	high
low	moderate	high	low	very high
low	moderate	high	low	very high
ow to mod.	moderate	fairly high	high	fairly high
ow to mod.	moderate	fairly high	high	—
moderate	moderate	moderate	moderate	—
—	low to mod.	mod. to high	low to mod.	—
moderate	moderate	moderate	moderate	high

there is a danger of local overcuring and degradation, as heat travels only with difficulty in hard rubber. This degradation may give rise to gas formation, and the consequent risk of explosion. Accelerators affect only the rate of vulcanization, they have no effect on the final properties.

Antioxidants are incorporated, but these protect the sulphur, not the rubber, preventing its loss by oxidation. Plasticizers and fillers are also incorporated for ease of processing. Fillers moreover prevent undue softening of the final product at elevated temperatures. One of the most widely used fillers is hard rubber dust, which recalls the use of reclaimed rubber in soft rubber mixes. Hard rubber dust—obtained by grinding waste hard rubber to powder on a special mill—reduces the danger of overcure and also produces very smooth surfaces on the final product.

The properties of hard rubber are those more often associated with plastics than with rubbers. It is a hard, tough material, with a high tensile strength and impact resistance. Although it is tough it can be machined easily. The chief disadvantage is its tendency to soften at high temperatures. It has good electrical resistance which makes it suitable for mounting switches. Its resistance to corrosion makes it a good lining for chemical plant storage tanks, and battery cases. Light-coloured vulcanites have been used for making the palate or plate part of false teeth, but this field has largely been taken over by plastics.

7

The manufacture of rubber articles

THE manufacture of rubber articles is the end of the line. Having obtained the rubber and treated, processed and tested it, the final operation is to make the article and sell it. Of course much of the processing already described is part of the manufacture, and all that is needed to finish many articles is to have the mould in the shape and dimensions of the required final product, and heat and vulcanization do the rest. On the other hand, some articles, for example golf balls, undergo as many as 40 different operations before the product is complete. As there are wide variations from one product to another, so there are frequently widely different ways of making the same thing.

For the purposes of this chapter only a representative selection of manufactured goods will be described and only the most usual way (as far as can be ascertained) of making them will be reported.

It was pointed out in Chapter 5, on the processing of S.B.R., that this synthetic rubber processed in a way akin to natural rubber, and this is true in a general way of most other synthetics. The manufacturing processes to be described will therefore apply to most synthetics, unless otherwise stated.

MECHANICALS

The term 'mechanicals' covers a wide range of moulded goods, and includes soles and heels, bathing caps, hot water bottles, cycle pedals, door-stops, tiles and other articles which form the component parts of many machines. Very often the articles are heavily loaded with fillers, and sometimes they contain reclaimed rubber to assist processing and reduce the cost. When the articles contain reinforcing fillers, S.B.R. can be used on equal terms with natural rubber and is often preferred because of its superior qualities. For example, in wringer rollers S.B.R. has a better service life as it is less affected by detergents and mild grease conditions than natural rubber. A particular field where S.B.R. has recently taken over from natural rubber is that of shoe-soling material.

FIG. 29. Soles being removed from a mould

Soles and heels

Since S.B.R. is less resilient than natural rubber it is often preferred for shoe-soling compounds where a less resilient material is more comfortable for the feet. Using the high styrene resins as reinforcing agents a further stiffening is obtained, and the result is very like natural leather and can be cut, sewn and worked generally in the same way as leather. The material will also stand flexing rather better than natural rubber and lasts longer than leather.

The mixing of the sole compound is carried out on a mill or in an internal mixer, and the finished rubber sheeted on an even-speed two-roll mill. 'Blanks', which in this case are usually rectangles of rubber, are then cut from the sheet and placed in preheated moulds. The moulds are in two pieces, and when the rubber has been inserted the two halves are held together in a hydraulic press. Heat is supplied by steam and curing takes from 10 to 15 minutes. The moulds are then opened and the soles removed. The mould cavities are usually sprayed with a lubricant before fresh blanks are inserted.

Fig. 30. Moulding the sole and heel to the upper

Fig. 31. The sole and heel now bonded to the upper

The cured soles will have a thin edge of 'flash' or 'spew'; rubber which has squeezed out between the halves of the mould during curing. This is trimmed off.

Heels are made in a very similar way. In this case, however, a little resilience is often desirable, and natural rubber is preferred to S.B.R. Sometimes instead of sheeting the rubber compound it is extruded as a long cord and sliced into blanks. Before the blanks are inserted, small steel washers are placed in the mould on pins. These will be the washers which, embedded in the finished heel, prevent the heads of nails from tearing through the rubber. When these have been placed in position the blanks are inserted and curing is effected in the same way as for soles.

A new process which is becoming widespread is the direct moulding of soles and heels to uppers of leather or canvas in one operation. The upper is fitted over a last, and tacked to a leather or fibre insole. The last is then withdrawn and replaced by one which forms part of a vulcanizing mould. A block of rubber to make some of the heel is placed in position and a thick strip of rubber compound, making the rest of the heel and the sole, is placed on top (see Fig. 30). The mould is then closed. Vulcanization takes a little over 10 minutes, and all that remains is the finishing operation.

Women's casual shoes, children's shoes, suède winter bootees and slippers having sponge rubber soles can all be made by this method.

Bathing caps

Bathing caps are made in a three-piece mould. A piece of metal in the shape of a head is covered by two halves, leaving a space in between. Two blanks are placed one on each side of the head, and the mould is closed in a press. Very short vulcanization times—of the order of three minutes—are usual for bathing caps, and a fast accelerator must therefore be used in the rubber mix. The edges of the cap are trimmed after vulcanization and the cap is turned inside out and is ready for use.

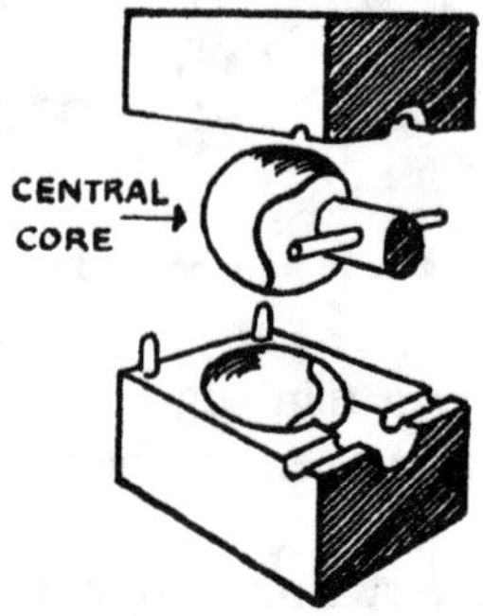

FIG. 32. Mould (diagrammatic) for bathing caps

Hot water bottles

Hot water bottles are another class of article made in a three-piece mould. In the same way as for bathing caps, a central core, which is the shape of the bottle, is surrounded by two enclosing pieces, and two blanks are placed one on each side of the core.

Very short cure times are again used and the mix has a low

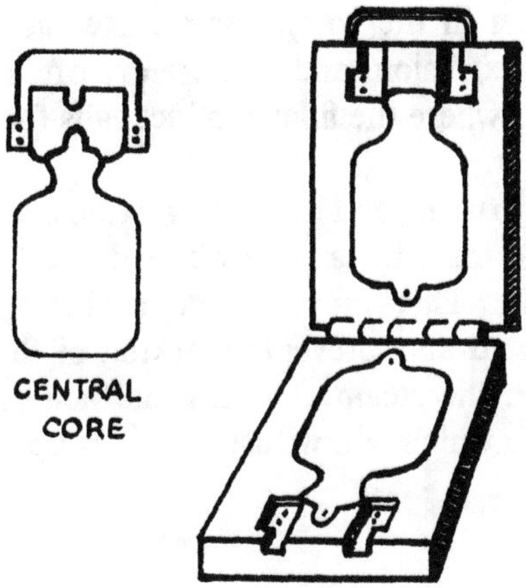

FIG. 33. Hot water bottle mould

sulphur content to increase its resistance to heat. When the moulds are opened the central core is removed with the rubber bottle around it. It is cooled and the rubber stripped from it. This involves stretching the neck to the width of the bottle to allow the core to come through, and care is needed or the rubber may tear. Because of this, and the use to which the bottle is put, a good elastic rubber is required with plenty of tear resistance. This means that natural rubber is generally preferred to the synthetics. The stopper and its holding neck must be a hard rubber, and this is sealed inside the bottle with a self-vulcanizing solution—containing a very fast accelerator which cures at room temperature. The finished article is trimmed and then tested by air inflation.

EXTRUDED ARTICLES

Like moulding, extrusion can provide very nearly finished products, or it may be an intermediate step. Examples of articles which are virtually finished after extruding and curing are brake-blocks, window-moulding, door-sealing sections and some forms of tubing. Rubber bands (stationers' bands) can be made by cutting an extruded and vulcanized tube, and of course inner tubes for tyres have only the valve fitted before vulcanizing.

Smooth extrusion of rubber is difficult to obtain, and complicated mixes are made to ensure good results. S.B.R. is less likely to scorch than natural rubber in the complicated mixes used and under the hot conditions of extrusion, and is therefore often preferred. Other synthetics are used where the final product calls for special qualities, as we shall see later.

For vulcanizing tubing and extruded sections, open steam is often used. The extruded material is merely coiled in a tray which is filled with French chalk. This supports the rubber, prevents it from sticking to itself, and also prevents marking of the rubber by condensed water from the steam. The tray is then placed in a vessel called an autoclave which will withstand pressure, and steam under pressure is admitted.

WIRES AND CABLES

Essentially, the element of any wire or cable system is a length of metal covered with insulating material. Normally, high elasticity is not required for the insulator, and for this reason many wires are plastic covered, but if used, a high quality rubber is demanded. The rubber must be a good electrical insulator, able to withstand abrasion and moderately high temperatures without breaking down, and to resist the deleterious effects of oxygen and ozone—which is often generated near electrical equipment. For these reasons some of the synthetics, particularly butyl, silicones and Neoprene have been used extensively for wire insulation in the U.S.A. for many years. Ap-

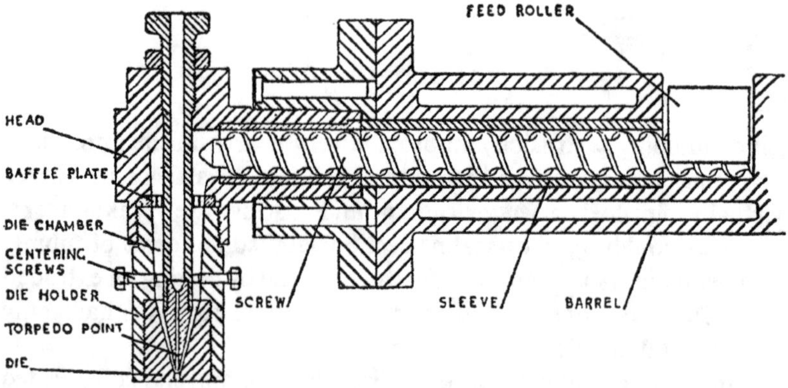

FIG. 34. Diagram of extruder for coating wire

proximately 75 per cent of the rubber used in the wire and cable industry there is synthetic.

The wire which is insulated is usually copper, though sometimes aluminium, silver or bronze are used. Copper wire is tinned by passing it first through a flux and then through a bath of the molten metal. The thin layer of tin keeps the metal of the wire and the rubber apart, so that the copper will not attack the rubber and the vulcanizing sulphur will not attack the copper. Tinning also enables electricians to strip the rubber off easily when making joints. To coat the wire with rubber a modified type of extruder is used. In effect it is an extruder with a nozzle bent at right angles to the barrel. The wire passes through a hole at the back of the nozzle and meets the extruded rubber coming along sideways. Both are then extruded together and leave at the nozzle, the wire covered with rubber. Usually the wire is pulled through the instrument by machine, and choosing a proper speed for this operation ensures a uniform coating of rubber.

HOSE

There are many different types of hoses made today. There are hoses to carry water, steam or compressed air, petrol, beer and various other liquids, and even solids like flour and wheat. Obviously the construction of these will be different and the materials used in making them will vary. For example, Neoprene and nitrile rubber are usually used to make hoses carrying petrol and oil, and butyl rubber is used for high pressure steam hoses. Nearly all these hoses will be reinforced with fabric, or metal and fabric, and it is in the reinforcement and the way in which they are cured that similarities appear.

Plain hose

Plain hose, without any reinforcement—apart from the chemical reinforcement of the rubber—may be of natural rubber or S.B.R. The hose is merely extruded and cured in open steam as with ordinary rubber tubing. As with rubber tubing, S.B.R. mixes show rather less tendency to scorch than natural rubber and, since it is less thermoplastic, hose tubes of S.B.R. hold their shape when hot rather better than those of natural rubber.

The fabric reinforced hoses may be split up into two types:

those that are reinforced with woven material and vulcanized in open steam, the so-called wrapped hose, and those that have braided reinforcement and are vulcanized in lead, the moulded hose.

Wrapped hose

In nearly all types of hose the first operation is to make a lining. This is usually a plain rubber tube extruded and placed on a mandrel. A mandrel is simply a long rod or sometimes a tube over which the hose is passed by blowing air between it and the mandrel. The woven material to be used in the reinforcement is next coated with rubber in a calender, and strips cut from the sheet slightly wider than the circumference of the tube to be covered. The strips are cut at an

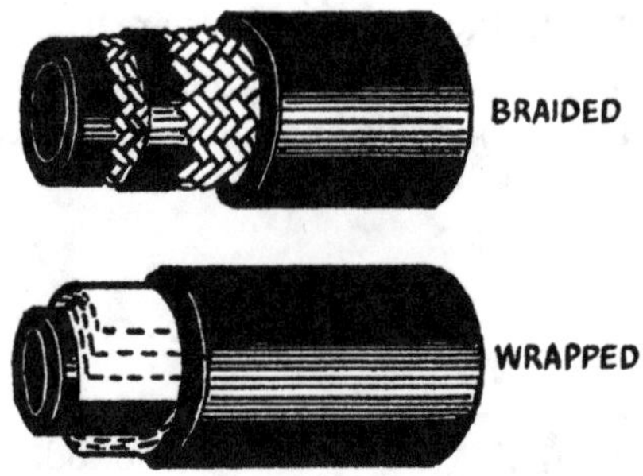

FIG. 35. Diagram of sectional hose

angle across the sheet—the bias angle it is usually called—so that when wrapped round the tube the threads in the woven material will not all be running in the same direction. This ensures that where the successive strips of material lie on top of each other the threads will not 'chafe' each other.

These strips are then joined together at their ends and wrapped round the lining tube by a special machine. One application of the reinforced material is called a ply. There may be more than one ply, but hoses of more than about four inches in diameter are usually built by hand. The outer casing of the hose is then applied. This is a calendered sheet of rubber and is applied in the same way as the fabric reinforcement.

The complete hose is then tightly wrapped with fabric to protect it from the steam and vulcanized in open steam in big autoclaves. When vulcanization is complete the mandrel is removed and the fabric is stripped off leaving a characteristic impression on the article.

Moulded hose

Wrapped hose is not made in lengths longer than about 60 feet, as longer lengths are too awkward to handle. Moulded hose can be made in almost any size or length, but it is not usually made in lengths of more than 500 to 600 feet. Most garden hoses when made of natural rubber are moulded.

As with wrapped hose the first step is to extrude the lining and this is then kept slightly inflated with the ends sealed. Braiding, which is made from a textile, for example cotton, is next applied to the lining by a special machine. The braided reinforced rubber is then passed through extruders similar in type to the one used for covering cables—that is with a right-angled nozzle—and a coating of rubber is placed on the braiding. Further plies of braiding may be applied and further rubber jackets until the required diameter is obtained. Then the hose is passed through yet another kind of extruder and a coating of hot lead is applied. This 'leadpress', as it is called, contains lead just below its melting point and under pressure. Under these conditions it becomes plastic and is easily extruded. The lead is cooled with cold water directly it leaves the press, and the hose coiled on a drum. The hose is filled with water and its ends clamped and vulcanization is effected with steam, the lead acting as a mould. After vulcanization the lead is stripped off for further use.

Moulded hose usually has a characteristic ribbed appearance, resulting from the nozzle of the lead press.

Armouring

Sometimes a spiral of steel wire is wrapped round the outside of a hose. This protects it if it has to be dragged along the ground and supports it if it is under high pressure. If the hose is a suction type, the armouring wire is usually inside the hose and prevents its collapse. It may be either along the bore or more usually between the plies.

There are two types of belts. Conveyor belts, which carry almost every conceivable type of material, and transmission belts, which vary from the small V belts driving the fan and dynamo on a car to large machine tool driving belts which may have a speed of up to 300 feet per second.

The rubber used in the construction of belts is black loaded, and its resistance to deformation, especially in transmission belts, is important. In transmission belting S.B.R. possesses no advantages over natural rubber and its use will be dictated only by cost. In conveyor belts, however, the superior abrasion resistance of S.B.R. makes it more desirable for the outside covering.

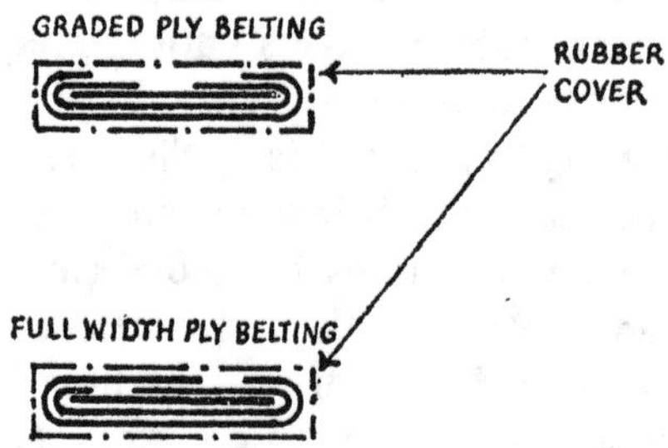

FIG. 36. Structure of plies in conveyor belting

Conveyor belts

Like hose, conveyor belts are made from rubber reinforced with fabric. The fabric may be cotton, nylon or terylene, or sometimes a mixture of textiles. It is thoroughly dried by being passed over heated rollers before the rubber is applied, and the still warm fabric is coated on a calender. Both sides of the fabric have to be coated, or, more accurately, 'frictioned', so that the material passes twice through a three-bowl calender, or once through two calenders in tandem.

The frictioned fabric is then cut and folded to make the required number of plies. There are several designs for different types of belts. Belts for carrying abrasive materials may have a graded ply system so that an extra thickness of rubber comes over the middle.

Building is done on long tables and care is taken that joints in the plies are not near each other. The joints in adjacent plies, for example, are staggered so that they are never less than 10 feet apart along the length. This ensures a more uniform thickness and the minimum of weak spots along the belt. After building, the whole belt is covered with a thick layer of calendered rubber and vulcanized.

Vulcanizing a long length of rubber is not easy. Sometimes it is vulcanized in sections in long presses, but this can lead to parts at the ends of the presses being cured twice, and today much belting is vulcanized continuously on a special machine, winding round a five-foot-diameter steam-heated drum.

Transmission belting

Transmission belts are made in a very similar way to conveyor belts except that the outer abrasion-resistant cover is often omitted. Normally transmission belts are much narrower than conveyor belts.

RUBBER BALLS

One of the earliest references made to natural rubber came from a Spanish historian describing the South American Indians and concerned their rubber play-balls, which bounced considerably higher than the European variety. The qualities described then are still valid today, and in many ways natural rubber is the best material to use for balls. The mix must give good tack, for the operations involve sticking several pieces of rubber together, and the final rubber must be resilient and stand repeated flexing.

Children's play-balls, tennis, squash and pelota balls are made in much the same way. The first operation is to produce a calendered sheet or sometimes an extruded tube. The next is to shrink the rubber to remove strains imposed by calendering and to ensure that the rubber will inflate evenly and not distort when vulcanized. This is done by warming the rubber either in hot water or on a hot plate.

When the shrinkage is complete, half-balls or shells are pressed out from the sheet on a special machine. The shells are then joined together in a mould. This mould closes before the shells are pressed together and is filled with air at about twice atmospheric pressure. The shells are then pressed together inside the mould and thus enclose air under pressure. Heat is applied and the rubber is half cured.

Finally, the balls are vulcanized in a two-piece mould. With tennis and other covered balls the covering is applied after vulcanization and is usually of woollen felt. The ball surface is first roughened and then the felt is stuck on with a fast vulcanizing solution. A few minutes in a steam-heated press is enough to stick the felt firmly on.

Golf balls

Unlike the other balls so far described, golf balls are not hollow. They are constructed about a central core. This core is in most cases two rubber cups moulded and sealed and usually vulcanized. Inside the core is a fairly thick and heavy paste which helps to give the final ball its weight and balance. The core is then frozen so that rubber tape can be stretched and wound round it. When a specified diameter has been obtained rubber thread is wound on, also under tension, until the ball has reached almost full size.

Nowadays the tape is usually wound on by hand, and the thread by machine. The rubber thread and tape are made from the same mix, which is a very simple one containing no fillers, and natural rubber is the main component. The rubber is accurately calendered and vulcanized in hot water.

The outer cover of the golf ball is made of gutta percha, a hard, tough material which is nevertheless chemically indistinguishable from natural rubber. Each half of the cover is stamped out separately and then placed over the core, and the halves are sealed on in a mould which also gives the cover its familiar indented pattern. After trimming, the ball is ready for painting.

Sometimes the covers are vulcanized, but not always. If they are to be vulcanized, very fast accelerators called ultra accelerators are used. Painting is the final operation.

TYRES

Nearly all tyres today, except industrial solid truck tyres, etc., are pneumatic; that is to say they are inflated with air, or contain an inner tube inflated with air. This type of tyre will therefore be the only one described.

Since half the amount of rubber used goes into tyres, the problem of how to use synthetic rubber for their manufacture is an important one. It would probably be true to say that nearly every improvement made in S.B.R. has been aimed at making it more suitable for tyres. Today S.B.R. competes on nearly equal terms with natural rubber in this field, which means that since S.B.R. has a more stable price than natural rubber and can compete economically, we must expect to see more of the world's tyre market changing to it. This applies particularly to the smaller sizes of tyre. With large truck tyres the

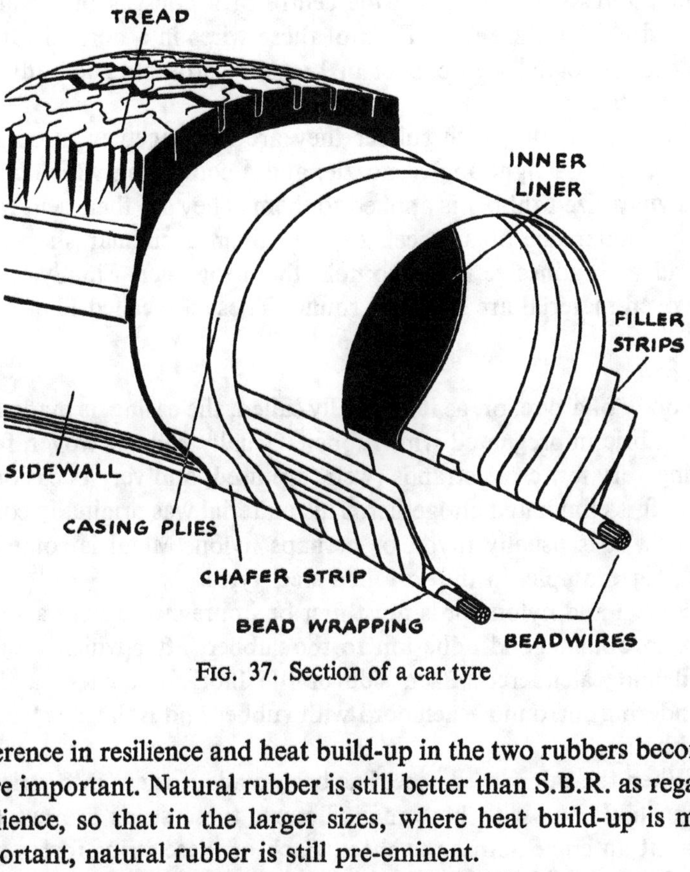

FIG. 37. Section of a car tyre

difference in resilience and heat build-up in the two rubbers becomes more important. Natural rubber is still better than S.B.R. as regards resilience, so that in the larger sizes, where heat build-up is more important, natural rubber is still pre-eminent.

The usual pneumatic tyre can be divided into two parts: the inner tube which carries the air, and the outer cover. Many modern tyres have no inner tubes, and the air is retained by the cover itself; but the construction of the cover is basically the same whether the tyre is tubeless or not, and the differences will be pointed out in the following description.

The cover

The cover illustrated in Fig. 37 is from a tubeless car tyre. Its many components can be split up into three sections: the bead, the cord fabric casing or carcase, and the tread.

The bead

The bead is the part of the tyre which holds it to the rim. It is a hard, rigid structure, as many people who have had occasion to remove tyres will know, and the centre of it consists of strands of steel wire. There are four to six of these wires in a normal car tyre and each is coated with copper or brass, to promote good adhesion to the rubber.

For the coating with rubber they are run together through an extruder with a right-angled nozzle, and a compound which gives a hard vulcanized rubber is applied to them. They are then bent round the circumference of a wheel, to give them a circular shape, and wound with adhesive fabric to hold them together. Finally lengths of woven material are wrapped round. These are called filler strips.

The casing

The body of a tyre, or, as it is usually called, the casing, is made from cord fabric impregnated with rubber. Usually rolls of woven fabric having very few cross strands (weft) are used, and very occasionally the weft is eliminated altogether. This material was originally cotton, but now it is usually rayon or perhaps nylon. Metal is sometimes used, for example, in the so-called steel tyres.

Rayon and nylon cords must first be impregnated with a special latex to obtain good adhesion to the rubber, after which they are dried and calendered. Each side of the fabric is coated, and the calendering surrounds each cord with rubber and isolates it from its neighbours.

When the calendered sheet has been made it is cut into strips on a long table. As with hose manufacture, these strips are cut on a bias—at an angle across the sheet. Each of these strips forms a ply and they are laid one on top of the other in the finished tyre, reversing the plies, so that the cords form a network. The bias determines the angle at which successive rows of cords will cross each other in the tyre. The angle is chosen so as to reach a compromise aimed to meet the varying requirements of comfort, inflation and driving stresses, and is usually about 45° in the finished tyre.

The tread and sidewalls

The tread is a thick layer of abrasion-resisting rubber, and the sidewalls are made of rubber with good resilience and flexing qualities. Each is made separately, and mixes are used which bring out the qualities required. The mixes are then extruded and the extruded strips carried on conveyor belts to a machine where heavy rollers press them together, the tread compound being in the middle and the sidewall compound on either side. The united strip is then cut into the required lengths, the cut being chamfered to ensure a strong seal when joined.

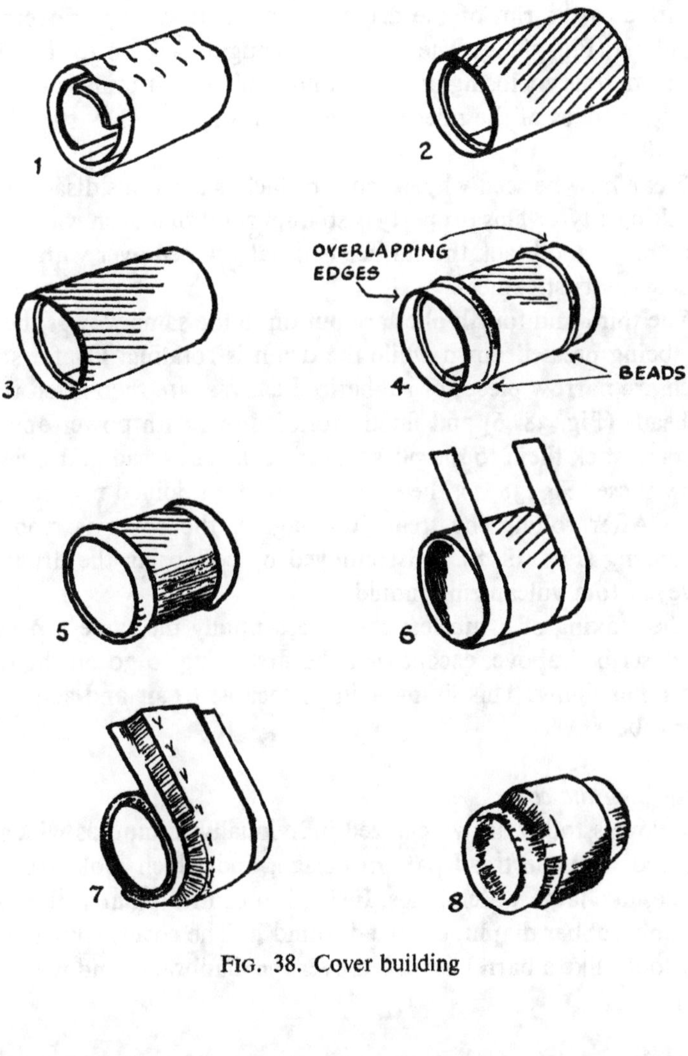

FIG. 38. Cover building

Cover building

Cover building is a skilled operation. The operator is like the man who carries the big bass drum in a band, except that the cover builder's drum is much smaller and is collapsible. Having collapsed his drum the operator takes one ply at a time of the fabric reinforced rubber and puts it round his drum (Fig. 38, 1). This is called 'single ply building'. Two or more plies can be assembled as a separate operation, and given to the operator to put on his drum. This is called 'pocket construction'. When the drum is pressed out to its proper size the plies are held firmly on the outside, with an overhanging edge (Fig. 38, 2 and 3). The bead is then brought up and fitted to the rim of the drum, where it sticks to the overhanging rubber (Fig. 38, 4). The rubber is brought back on to the top of the drum, thus enclosing the bead, and while the drum is rotated the folded-up edges of the plies are pressed down until they stick firmly (Fig. 38, 5).

It can now be seen why absence of tack is a serious disadvantage in making a tyre. This property is so important that even with rubber which has a good tack the surface is usually wiped over with solvent to increase its stickiness.

The third and fourth plies are put on in the same way as the first two, being pressed down while the drum is rotating. Chafer strips, which are narrow pieces of rubberized canvas, are then fitted round the beads (Fig. 38, 6) and finally rolled down with power-operated rollers to stick them to the plies underneath. The tread and sidewalls follow these (Fig. 38, 7), the strip is wound on, joined up and rolled down. After rolling the tread joint again, the cover is ready for vulcanizing (Fig. 38, 8). It is removed by collapsing the drum and conveyed to a vulcanizing mould.

The making of a tubeless tyre is essentially the same process as that described above, except that the first thing to go on the drum is an inner lining. This lining is impermeable to air and acts as the inner tube.

Vulcanizing the cover

Most covers today are vulcanized individually in moulds which are engraved with the tread pattern design, and which look, from the outside, like large watch-cases. Inside is a central pillar with a thick inflatable rubber diaphragm fixed round it. The cover, which at this stage looks like a barrel, is placed over the diaphragm and the mould closed. Compressed steam is then passed into the diaphragm and when inflated this forces the cover against the mould. The mould is heated by a steam jacket and vulcanization takes place, the steam in the expanded diaphragm supplying heat for the inside of the cover.

The diaphragm, or bladder, is an important feature of tyre-making, and its cost is an important factor in the price of the tyre. It has to be strong, for it supports a steam pressure of several hundred pounds per square inch, and it has to be extensible and able to expand and force the cover against the mould. Most important of all, it must stand these conditions and high temperatures many times a day. Owing to its very good ageing qualities most bladders today are made of butyl rubber.

Vulcanization of a normal car tyre takes about 30 minutes, after which the bladder is deflated and the cover removed from the mould. The flash and spew are trimmed off, and after inspection the tyre is wrapped for transporting.

Inner tubes

Most inner tubes today are made from butyl rubber because of its very low air permeability. After compounding, the butyl mix is extruded and cut to the required length. The valve is stuck on and the ends of the tube joined by cutting a piece off each end with a hot knife to expose a fresh surface, and bringing them up to each other quickly so that they stick together. This operation is called 'butt jointing' and is carried out by machine.

The tubes are vulcanized in moulds of a kind similar to those used for tyre covers. Air is blown into the tubes and this forces them against the moulds during vulcanization, when the butt joint disappears.

GOODS FROM LATEX

Most rubbers today, synthetic and natural, are produced in the form of a latex. A latex is simply an emulsion of the rubber in water, and is similar in many respects to milk, which is an emulsion of butter-fat in water. Natural rubber comes from the rubber tree as a latex and many synthetic rubbers (since their formation is by emulsion polymerization) appear as a latex.

Rubber hydrocarbon constitutes between 30 and 35 per cent of the weight of natural rubber latex. Resins, proteins, sugars and salts make up between 4 and 5 per cent, and the rest, 60 per cent, is water. Since natural rubber latex very easily becomes infected with bacteria, it will not keep long after being removed from the tree, and it was not until efficient preservatives were found that it became of much practical use. Even then the fact that two-thirds of the latex is water, and that it had to be shipped all the way from the Far East, meant that it was not a very economic raw material to use in manufacturing.

This disadvantage has been overcome to a certain extent by concentrating the latex before shipment. This can be done in a number of ways of which centrifuging, creaming and evaporation are all important and in common use.

The same difficulty is met with in synthetic latices, although usually they do not have to be transported so far. It is, however, often more convenient to concentrate the latex, and similar means are used as for natural rubber latex.

But it cannot at the moment be said that synthetic rubber latices have found such a wide application as the synthetic rubbers themselves. Films from S.B.R. latex have nothing like the strength of films from natural rubber. At present S.B.R. latex finds most of its applications where support can be given to the rubber film, for example, in the treatment of textiles or in tyre cord dipping. Most of the applications to be described here, therefore, will apply to natural rubber latices, except for some dipped goods where Neoprene latex is chosen for its special properties of heat and oil resistance.

Compounding latex

Compounding latex is rather different from compounding masticated rubber. To begin with, in natural rubber latex not so many compounding ingredients are added as in the dried rubber. Those that are added include zinc oxide, sulphur, an accelerator and sometimes a colouring agent. The accelerator is usually a very fast one, giving a cure in a few minutes. The ingredients are usually made into a dispersion with water by grinding them up in water, and the dispersion is then simply stirred into the latex.

If the latex is then warmed vulcanization will take place while the rubber is still in the emulsion form. The latex looks no different, but on drying down it will be found that the rubber has in fact been vulcanized. However, a vulcanized latex is not always suitable for making products, particularly latex foam, and for some applications what is called a vulcanizable, that is simply a compounded, latex is used in preference to a vulcanized one.

Dipped goods

A wide variety of rubber goods can be made by dipping. These include feeding-bottle teats which are transparent, and opaque articles such as balloons and gloves, fountain-pen sacks, bottle caps, bathing caps and rubber boots.

The latex to be used for dipping is strained through muslin and placed in a tank which is itself surrounded by another tank containing water to keep the temperature constant. The latex is left for about 30 hours before any dipping begins. A former, which is usually of porcelain, is then dipped both into the latex and into a coagulating material which sets the deposit immediately. Most products are built up in a series of rubber layers by alternate dipping into the latex and then into the coagulating tank. In this way household gloves, bathing caps and the outside layers of rubber boots are made. The formers are washed and the article is stripped off, often under water, washed again and dried. Vulcanization is effected in a large hot-air oven.

FIG. 39. Dipping rubber boots

Latex thread

An extension of the dipping technique is used in making rubber thread from latex. Compounded 60 per cent latex is used for the purpose, and of the synthetic latices only Neoprene has been used with much success. The latex is compounded with zinc oxide and possibly another white filler, sulphur, accelerator, antioxidant, and sometimes a wax. It is carefully strained before use to remove coarse particles, and then squirted under pressure through glass nozzles which lie just beneath the surface of the coagulating tank. Coagulation occurs immediately and the round thread is pulled for 10 feet or more through the tank. After this it passes through a washing tank. It is dried and partially vulcanized by being passed through an oven, dusted with French chalk and rolled up on a spool. Vulcanization is completed on the spool by heating for a few hours at a low temperature. This ensures an even cure throughout the spool.

Latex thread may be made in a number of thicknesses, depending on the size of the nozzle, the pressure, how fast it is pulled through and so on. The thinner grades are nearly always wrapped with threads of textile materials—cotton, rayon, nylon, etc.—which are wound on by a special machine and are then known as elastic yarns. The thinner grades find applications in knitting and hosiery or as the weft in finely woven cloths, and are usually sold in hanks, or wound on cones.

The thicker thread is usually called simply latex or rubber thread, and is often used without any textile wrapping. It is used in three ways: making elastic web, elastic braid and elastic cord. Elastic web is used in braces, suspenders and belts, and in it the rubber thread is interwoven with a textile material. Elastic braids, more familiarly known as 'knicker elastic', contain rubber thread braided on a special machine with a textile. Elastic cords may be a single strand of heavy gauge rubber, or a group of threads bound together with a textile covering. The former is used in hat elastic, and the latter as springs.

Latex foam

The largest use of latex today, either natural or synthetic, is in making latex foam. Mattresses, pillows, seat cushions for all types of seats, carpet underlays and even toys can all be made from latex foam. Polyurethane foam has been described briefly in the last chapter; the following applies only to natural rubber foam.

Sponge rubber has been made from masticated natural rubber for many years, but it was not until latex was employed that its use became widespread. Substitution had had to await adequate supplies of good quality concentrated latex, which appeared in the late 1920's.

The process most used today is called the Dunlop process, since it was invented in the laboratories of the Dunlop Rubber Company. It consists of adding a slight excess of soap, over and above that needed to stabilize the latex, and of beating air into the latex with a whisk. The whisk may be very like those used for cake mixing. The latex froths up to something like 10 times its original volume, depending on how much air is beaten in. The more air the lighter the foam and the softer the final product.

In the later stages of foaming a delayed action coagulant is added, and when the mixture has the consistency required it is poured quickly into a mould. The latex then coagulates and at the same time the foam structure changes. The individual air spaces become interconnecting. Vulcanization takes place in the same mould and is effected by steam. The mould is then opened and the steaming article removed, washed through in water and dried. Much of the water can be squeezed out by hand; this is supplemented by centrifuging the article, which removes more water, and finally by drying in hot-air chambers.

The foam rubber thus made has a permanent structure of tiny interconnected air spaces so that the rubber is porous. This makes it particularly comfortable material for cushions, as well as making it hygienic.

There are many other applications of latex, but space forbids their inclusion. A section on the uses of latex, however, would not be complete without a mention of chewing gum. Although alleged to be more popular in America, it still has many devotees in this country. Originally chicle, a low-grade Central American rubber, was used exclusively, but today S.B.R. latex has largely replaced this, except perhaps for bubble gum, which is of a firmer type and requires the use of natural rubber latex.

PART THREE

The Future of Rubber

8

Modern developments

THIS chapter is concerned with modern developments, and since these are made in various sections of the rubber industry, the chapter will be something of a mixture. Possibly the most important development has been the making of a rubber which is virtually identical to natural rubber. In fact one can say that natural rubber has finally been synthesized. Thus the wheel has turned full circle, for we began with natural rubber, passed on to synthetic rubber, and now finally arrive at synthetic natural rubber.

As we saw in the chapter on special purpose rubbers, modern ideas have gone beyond natural rubber. Some synthetics have properties which put them in new fields, and the tendency today is to think not so much in terms of general purpose rubbers like natural rubber and S.B.R., but of special purpose rubbers, each one being tailored to satisfy some particular need. The making of synthetic natural rubber therefore may not be such a world-shaking event as it looks at first sight. Indeed other developments may turn out in the long run to be more important.

Nevertheless the new synthetic natural rubber will be described first and some other developments later. To begin with, it is proposed to deal a little more fully with some of the chemistry of polymerization, and to lead from this into synthetic natural rubber. The reader whose interest does not lie in this direction is therefore advised to skip the next few sections and start with synthetic natural rubber proper.

The chemical reaction known as polymerization, which is the joining up of small molecules (monomers) to form large ones (polymers), can proceed in two different ways. One way is called 'condensation polymerization', the other 'addition polymerization'.

Condensation polymerization

In condensation polymerization, two molecules react together and join up, and another smaller molecule is eliminated or condensed between them. Fig. 40 illustrates this.

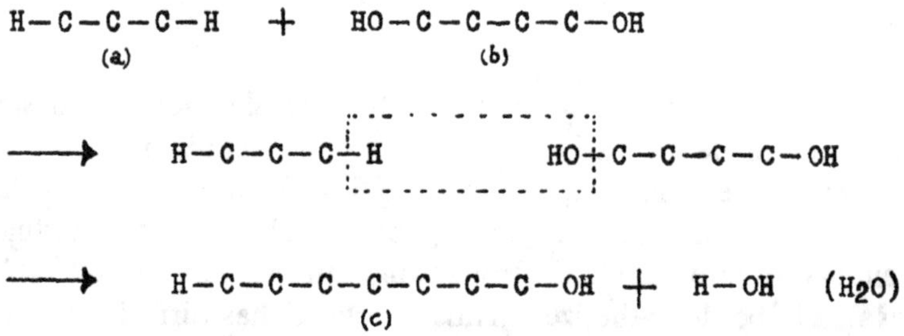

FIG. 40. Diagrammatic representation of condensation polymerization

Molecules *a* and *b* come together, then a hydrogen atom from *a* and an OH group from *b* is removed to make the molecule HOH (H_2O), which is water. The remaining parts of the molecules join up to make the long chain *c*. Since the final molecule *c* has an OH group at one end and a hydrogen atom at the other, both ends may add on further molecules of *a* and *b* respectively, thus building up a longer chain. This type of polymerization is the one which gives rise to polysulphide rubbers and silicones, and in some cases polyurethanes.

Addition polymerization

In addition polymerization the monomer molecules merely add on to each other, and no atoms are eliminated or condensed between them. Normally an initiator is used in the reaction. A molecule of initiator—or sometimes only part of the initiator molecule—first attacks a molecule of monomer. Attack usually takes place at the double bond in the monomer molecule. This is changed to a single bond and the initiator adds on to the monomer. The breaking of the double bond makes the monomer molecule very reactive and the combined monomer and initiator attacks another monomer molecule and adds it on. The new monomer molecule is activated —having lost a double bond—and the process of addition can go on, theoretically, until all the monomer molecules are used up. But the chain may stop growing in various ways. Two active chain ends may join up and then each loses its activity. Or another initiator molecule may be joined on and this also stops the chain growing.

In condensation polymerization the monomer molecules usually only add on to the polymer at its ends, for example where the hydrogen atoms or the OH groups are. But in addition polymerization the addition of a monomer need not take place at the ends of a polymer chain, and often does not, so that branches may be formed.

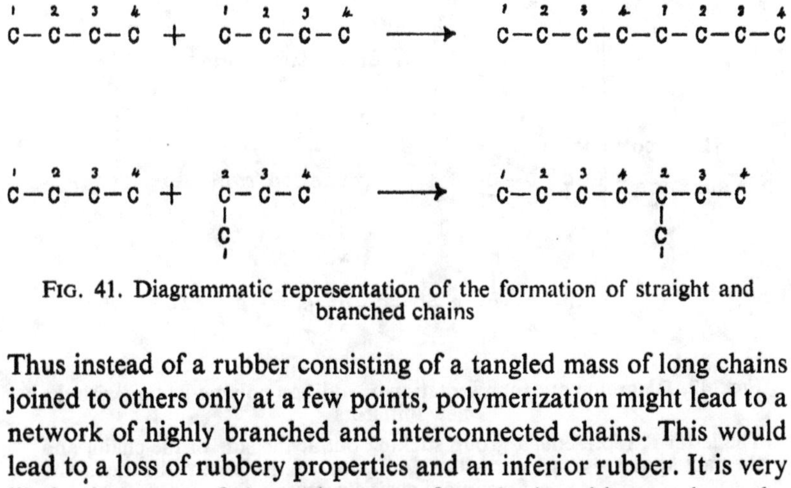

FIG. 41. Diagrammatic representation of the formation of straight and branched chains

Thus instead of a rubber consisting of a tangled mass of long chains joined to others only at a few points, polymerization might lead to a network of highly branched and interconnected chains. This would lead to a loss of rubbery properties and an inferior rubber. It is very likely that most of the early types of synthetic rubber, at least the general purpose types, suffered from this disadvantage.

It seems that there are two factors which influence the structure of the final rubber. One is the temperature at which polymerization is conducted and the other is the initiator which enables polymerization to proceed. By and large, the lower the temperature used for polymerization the better the product is, with less branching of the chains and more end-to-end addition of monomer units. This is why most S.B.R. is now polymerized at a low temperature, and also why butyl is polymerized at an even lower temperature. But more important than the temperature is the type of initiator used.

Initiators

There are two types of initiators in common use today. One is called the free radical initiator and the other the ionic type. The free radical type of initiator is the one commonly used in emulsion polymerization, that is for S.B.R., nitrile and Neoprene rubbers. Recently some new ionic initiators have been discovered. The really important development has been the introduction of these, rather than the production of synthetic natural rubber. They were first patented by a German chemist, Karl Ziegler, in 1953. Later they were used to polymerize isoprene and butadiene.

These initiators appear to work differently from those we have

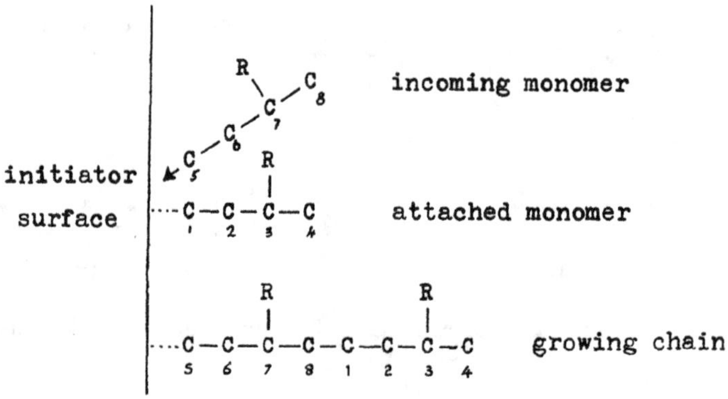

FIG. 42. Diagrammatic representation of addition polymerization with the new initiators

The letter R represents a group sticking out at the side of the chain, and the fact that these appear all on one side of the chain illustrates the forming of a *cis* structure

met before. Polymerization begins when a molecule of monomer comes into contact with the initiator. In free radical polymerization the initiator attacks one end of the monomer molecule and the other end usually becomes active and can add on further monomer units. With the new ionic initiators, one end of the monomer molecule is attacked as before, but when further monomer units are added the chain grows from the initiator end, not the other. Now it appears that with these initiators a monomer molecule, to be added at all, must be inserted in a special way. The result of all this is that a chain is formed with practically no branches.

But more than this, a chain is made of a particular structure. To understand this we must look a little more closely into the arrangement of atoms in a chain.

The arrangement of atoms in a chain

The diagrammatic representation of the atoms in a section of rubber or polyisoprene chain, first given in Chapter 2 and reproduced here in Fig. 43*a*, is a shortened way of writing it and does not represent the true state of affairs. In the first place the chain is not a straight one of carbon atoms as shown. Each bond between carbon atoms is actually at an angle to the next one, and a better representation is a sort of zig-zag pattern. The double bonds also introduce a complication, and on the whole it is better to show the double bonds as straight and the other bonds sticking out at angles. It should be borne in mind that the chain is a three-dimensional structure and is thus difficult to represent in two dimensions.

The sum total is that there are two ways of writing the polyisoprene chain, and each represents a definite structure different from the other, (see Fig. 43*b* and Fig. 43*c*).

In one way, called the *cis* configuration, all the CH_2 groups in the molecule lie on one side of the chain, and the CH_3 groups and one hydrogen atom lie on the other side. In the *trans* configuration, pairs of CH_2 groups lie on alternate sides of the chain and single CH_3 groups and hydrogen atoms lie on opposite sides of the chain.

Now this seems a small difference, and of course it is. But it is a definite difference, and if the chain were made up of coloured balls joined by rods, as is sometimes seen in models, no amount of turning the chain about would make the two forms the same. The startling thing is that this slightly different arrangement of atoms makes all the difference between rubbery and non-rubbery properties, for the *cis* form is natural rubber and the *trans* form is gutta percha, the hard waxy solid used to make the outside covering of golf balls.

This sort of difference, which is called chemically 'isomerism', can occur in any molecule which contains double bonds. It can occur in polymers made from butadiene and dimethyl butadiene, and they can therefore exist in two forms, *cis* and *trans*. Now we see why

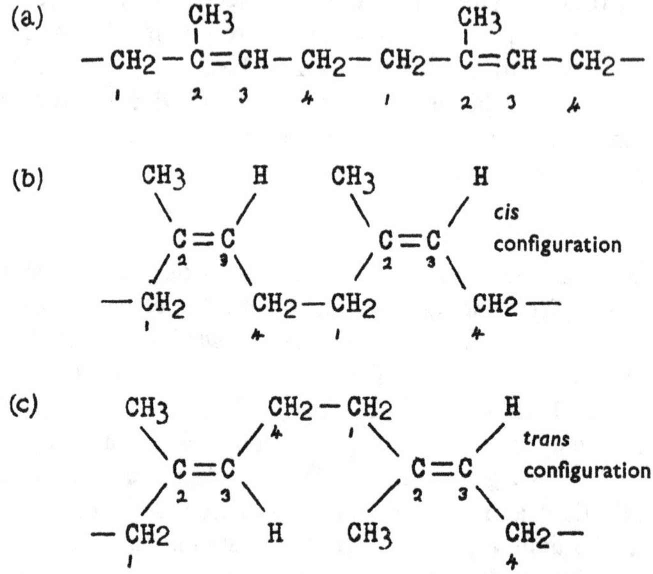

FIG. 43. Polyisoprene chain

many of the early synthetic rubbers were not much like natural rubber. For natural rubber is the *cis* form almost entirely, without branching, and many of the early synthetic rubbers were mixtures of *cis* and *trans* forms, also having heavily branched chains.

By making small changes in the initiator one is able to change the product of polymerization. With one particular initiator very nearly all *cis* will be obtained, with another very nearly all *trans*, and with yet another both *cis* and *trans* forms will occur. Thus the polymer can be changed at will. It is truly tailor-made.

SYNTHETIC NATURAL RUBBER

In December 1954, the president of B. F. Goodrich, the American rubber company, announced that members of the research group working with those of Gulf Chemicals had made a rubber substantially the same as natural rubber. There was an impressive weight of chemical and physical analytical evidence to support this. Some more information was given in November 1955 when it was disclosed that information on initiators had been purchased from

Dr. Karl Ziegler of Germany. At the same time the Firestone Rubber Company of America announced that they too had previously prepared synthetic natural rubber using a different initiator, and they were followed by Goodyear who had done the same thing.

These rubbers were substantially the *cis* form and contained no branching, and were very similar to natural rubber in all their properties. Perhaps it was too much to expect that they would be identical to natural rubber, for it must be remembered that natural rubber is a plant product and a complex mixture, containing not only the rubber hydrocarbon but protein, resins and salts as well. The new synthetic rubber needed little mastication, and substances could be mixed into it in very much the same way as with natural rubber. Antioxidants were needed and were added as soon as the rubber was made, in the same way as for S.B.R.

The rubber had excellent tack. The tensile strength, without reinforcing fillers, was as high as that of natural rubber, and in nearly every way it was just about the equal of natural rubber. In some respects, for example in its reaction to oxygen, synthetic natural rubber proved to be better than natural rubber, and it also had rather higher thermal stability. Tests were soon made on tyres and the tread wear was found not quite as good as natural rubber, but the resilience and heat build-up were as good, and of course much better than S.B.R.

This at once raised the question of tyres for heavy duty on lorries, the field in which natural rubber has been pre-eminent. The first results were encouraging. The United States Government recently tested synthetic natural rubber in military tyres and found it an adequate replacement for natural rubber. Goodyear also conducted extensive tests with synthetic natural rubber for lorry tyres and reported success. It looks, therefore, as if synthetic natural rubber can be an effective replacement for natural rubber, provided it can be economically produced.

Synthetic natural rubber cannot be produced in current S.B.R. plants, and entirely new plants will have to be built for its commercial exploitation. It is not emulsion polymerized, and much of the technology is therefore different. One or two of the American rubber firms, however, are building plants to produce it, and Goodyear have in mind the production of 25,000 to 30,000 tons a year. It appears that the principal difficulty is the monomer isoprene, which is at present being produced in inadequate amounts.

But if isoprene can be produced cheaply and in quantity, synthetic natural rubber may be an important factor in the future.

cis POLYBUTADIENE

Butadiene, like isoprene, gives a polymer which contains double bonds, and the same forms are applicable to polybutadiene as to synthetic natural rubber (polyisoprene). *Cis* and *trans* forms have been prepared using similar types of initiator to those used for polyisoprene, and the *trans* form emerges as a fibrous material, while the *cis* form is a rubber. It appears to be quite a good rubber with better low temperature properties than natural rubber or synthetic natural rubber.

It has been suggested that *cis* polybutadiene could be blended with natural rubber without reducing the properties of either too much. This is an attractive idea as butadiene is much cheaper than isoprene at present, and abundant supplies of it are available. Polybutadiene is therefore likely to be an economic proposition, and even if all its properties do not equal those of natural rubber or synthetic natural rubber, it seems likely to find many uses.

RADIATION AND RUBBER

In 1896, a few years after Tilden had polymerized isoprene, the French physicist Becquerel noticed that a uranium compound fogged a photographic plate when placed near it, even though the plate was wrapped in black paper. This, he demonstrated, was because the uranium compound gave off radiation which passed through the paper and affected the plate. The activity was called radioactivity, and later quite a number of substances were shown to exhibit it. In the following years many things were learned about radioactivity, not the least important being that it could cause many chemical reactions to take place. The fogging of the photographic plate was the first example, for this does involve a chemical reaction. But not much practical use could be made of the new tool as the quantity of radioactive substances available was so small.

This scarcity has changed in the last few years. During and since the last war atomic piles have been built, and these are sources of very penetrating and powerful radiation as well as producing many substances (radioisotopes) which are themselves radioactive, so today there is a substantial source of radiation which can be turned to chemical use. What can be done with it? The types of reaction which radiation can induce, and which might be useful in polymer science, can be grouped under a number of headings. The first is polymerization.

Polymerization

Some monomers can be polymerized merely by irradiation. How this occurs is not quite clear, but it seems that radiation can take the place of initiators. This type of polymerization has some advantage over the chemically induced reaction, for there is no initiator residue to remove when the reaction is complete, and radiation appears to be equally effective over a range of temperatures, which can therefore be adjusted to give the best product.

Cross-linking

It is possible to cross-link or vulcanize some polymers by radiation, and once again no other agency need be employed. The amount of cross-linking can be controlled by the amount of radiation put in, and the curing process is uniform and can be carried out at low temperatures so that the rubber is not subjected to excessive heating. Since no sulphur has been used for vulcanization, the chain cannot have sulphur cross-links, and the properties of the radiation-cured rubber are a little different from sulphur-cured rubber. For example, while the resilience is usually good, the tensile strength is low.

The fact that radiation curing requires no external heating is not always an advantage, for with moulded goods some heat is necessary to soften the rubber so that the blank can fill the mould before vulcanization hardens it. To produce moulded goods by radiation it seems that a very soft or even liquid compound may be necessary, or, alternatively, some heating will be needed to ensure proper moulding. There are, in fact, quite a number of drawbacks to the use of radiation. In some cases, as for example in butyl rubber, it does not induce cross-linking at all, but degrades the rubber.

The main factor affecting the use of radiation is economics. The atomic reactors available in this country today, apart from the experimental ones, are all designed to raise steam for turning turbines to make electricity, and these cannot be used for curing or initiating polymerization. But what are called the spent fuel elements —the fuel elements after the plutonium has been extracted—can be used. In fact tyres have been cured by the use of these elements. They cannot, however, be moved far from the atomic pile, and the radiation they emit, while very intense at first, falls off rapidly after a time. To make use of this source, therefore, manufacturers would have to consider the economics of building a plant near an atomic pile and using a variable source of energy.

There are other sources of radiation, however. The electrical machines known as particle accelerators produce radiation which may be suitable for some applications. It is not very penetrating radiation, and the cost of these machines is high, running into many thousands of pounds, but if a process could be designed so that very penetrating radiation was not needed and these machines were in fair demand, the cost to build them might be lowered.

Another source of radiation is radio isotopes, which are radioactive materials made in atomic piles. These give very penetrating radiation and the materials, as with fuel elements, have to be surrounded by thick walls of concrete. The cost of using radio isotopes is also high, but most research laboratories using radiation as an experimental tool employ them as a source.

The most useful outlet so far for radiation in the polymer field has occurred in plastics, where the cross-linking of polyethylene has given it a much higher softening temperature. But the effective cross-linking of most rubbers by radiation is much more expensive than by normal curing systems. The cross-linking of silicone rubber does not seem to be as expensive as many of the others, and since silicone rubbers are not cheap anyway, there might be commercial possibilities in cross-linking these rubbers by radiation.

It is difficult to forecast the future. New additives might be developed which would bring the high cost of cross-linking by radiation down, or allow advantage to be taken of the fact that radiation does not require heat.

At present polymerization is effected by radiation where chemical means fail. There is much research being done on this topic at the moment and the next few years might see some worthwhile developments.

LOOKING AHEAD

It is no easy task to conclude this book. I have the feeling that there are a great many things which ought to be included: new rubbers, additives, and processes to improve the final product. A halt must, however, be called somewhere. Since scientific prophecy is always a risky business it seems to me that the best way to end will be to describe briefly the use of rubber in what is perhaps the most forward-looking instrument of modern times—the rocket.

There are two types of rockets in use today, liquid propellant and solid propellant, and rubber in one form or another is often used in both. The liquid propellant type of rocket is more complicated in design than the solid propellant one. Briefly, two liquids, liquid oxygen and alcohol for example, are sprayed into a combustion chamber where they meet and react together vigorously. The hot gases evolved are ejected from the back of the rocket, thus pushing it forward. The liquids are stored separately in the rocket and pumped to the combustion chamber when required.

Rubber is often used in gaskets, O rings and seals in this type of rocket, and butyl, silicone and Neoprene rubbers have been chosen because of the extreme conditions of heat, solvent attack and pressure which the components have to withstand. In this type of rocket, too, shock resistant mountings in the controlling mechanisms are of rubber.

The solid propellant rockets are less complicated than the liquid fuel ones. The solid propellant is held in a case at the back of the rocket. The propellant burns in the case and as before the hot gases are ejected through a nozzle.

Solid fuels are usually an intimate mixture of what is called a fuel binder, and something to burn or oxidize the fuel, an oxidizer. The fuel binder is often a rubber, either a soft polymer type like Thiokol or one with a lot of plasticizers in it to make it soft.

The oxidizer is a solid and this is first ground to a fine powder, then mixed with the rubber in a special mixer. Carbon black, antioxidants and curing agents are also added. This is an extremely dangerous operation because of the risk of fire, and the mixer is usually sited in a room with thick concrete walls and no one goes near it while mixing is in progress.

The propellant mix is extruded to form a billet and the outside is coated with a fire-proofing material. The billet is then inserted in the case and cured. This is the procedure if the propellant is meant to burn at the end only, like a cigarette. If it is meant to burn throughout its length, and this is more common since it gives a more even burning and therefore a steadier thrust, it has a hole down the middle. Usually it is a star-shaped hole, and this is obtained by inserting a mandrel of this shape in the case, pouring the mix round it and curing. The mandrel is then removed.

Solid fuel rockets have less flexibility than the liquid fuel types. They cannot be stopped once started, and they cannot be made quite as big as liquid fuel rockets. Although liquid fuel types do not require rubber in the fuel, they do require it in small but vital accessories. It is a fitting thought to end this book, then, that every rocket launched today depends to some degree or other on rubber.

Bibliography

This book list has been split up into subjects or topics, so that the reader requiring further information on a particular subject can find it easily. As many of the books cover more than one subject, classification is difficult, and a book has been placed, therefore, in the category which seems to fit as large a proportion of its contents as possible.

HISTORY

HAYNES, W. 1945 — *This Chemical Age*. Secker and Warburg, London

McFADYEAN, SIR A. 1944 — *The History of Rubber Regulation 1934-1943*. George Allen and Unwin, London

SCHIDROWITZ, P. AND DAWSON, T. R. 1952 — *History of the Rubber Industry*. W. Heffer and Sons, Cambridge

WOODRUFF, W. 1958 — *The Rise of the British Rubber Industry*. Liverpool University Press

PHYSICS AND CHEMISTRY

DAVIES, C. C. AND BLAKE, J. T. 1937 — *The Chemistry and Technology of Rubber*. Reinhold Publishing Corporation, New York

FISHER, H. L. 1957 — *Chemistry of Natural and Synthetic Rubbers*. Chapman and Hall, London

LE BRAS, J. 1957 — *Rubber: Fundamentals of its Science and Technology*. Crosby Lockwood and Sons, London

MARK, H. AND WHITBY, G. S. 1946 — *Advances in Colloid Science*, Vol. II. Interscience Publishers Inc., New York

MEMMLER, K. 1934 — *The Science of Rubber*. Reinhold Publishing Corporation, New York

TRELOAR, L. R. G. 1958 — *The Physics of Rubber Elasticity*. Clarendon Press, Oxford

WHITBY, G. S. 1954 — *Synthetic Rubber*. Chapman and Hall, London

COMPOUNDING, TECHNOLOGY AND TESTING

BARRON, H. 1947 — *Modern Rubber Chemistry*. Hutchinson's Scientific and Technical Publications, London

COOK, P. G. 1956 — *Latex Natural and Synthetic*. Chapman and Hall, London

FALCONER FLINT, C. 1938 — *The Chemistry and Technology of Rubber Latex*. Chapman and Hall, London

MOAKES, R. C. W. AND WAKE, W. C. 1951 — *Rubber Technology*. Butterworths Scientific Publications, London

NAUNTON, W. J. S. 1954 — *What Every Engineer Should Know About Rubber*. The British Rubber Development Board, London

STERN, H. J. 1954 — *Rubber: Natural and Synthetic*. Maclaren and Sons, London

WHITBY, G. S. 1920 — *Plantation Rubber and the Testing of Rubber*. Longmans, Green and Co., London

WINSPEAR, G. G. 1958 — *The Vanderbilt Rubber Handbook*. R. T. Vanderbilt Co., New York

SPECIAL PURPOSE RUBBERS

DOMBROW, B. A. 1957 — *Polyurethanes*. Chapman and Hall, London

HAUSNER, H. H. 1958 — *Modern Materials*, Vol. I. Academic Press Inc., London

McGREGOR, R. R. 1954 — *Silicones and their Uses*. McGraw-Hill Publishing Co., London

ROCHOW, E. G. 1951 — *An Introduction to the Chemistry of the Silicones*. Chapman and Hall, London

POLYMERIZATION

BAWN, C. E. H. 1948 — *The Chemistry of High Polymers*. Butterworths Scientific Publications, London

BILLMEYER, F. W. 1957 — *Textbook of Polymer Chemistry*. Interscience Publishers, London

D'ALELIO, G. F. 1952 — *Fundamental Principles of Polymerization*. Chapman and Hall, London

FLORY, P. J. 1953 — *Principles of Polymer Chemistry*. Cornell University Press, New York

MARCHIONNA, F. 1946 — *Butalastic Polymers*. Reinhold Publishing Corporation, New York

High Polymers—a series of Monographs on the Chemistry, Physics and Technology of High Polymeric Substances. Interscience Publishers, Ltd. 88-90 Chancery Lane, London, W.C.2. (Catalogues on request.)

MISCELLANEOUS

BOVEY, F. A. 1958 — *The Effects of Ionizing Radiation on Natural and Synthetic High Polymers*. Interscience Publishers Ltd., London

WARREN, F. A. 1958 — *Rocket Propellants*. Chapman and Hall, London

Glossary

Abrasion: The ease with which rubber wears away.

Accelerators: Substances which accelerate the vulcanization of rubber, enable less sulphur to be used, and sometimes enable the vulcanization to take place at a lower temperature.

Acetone: A colourless inflammable liquid, used as a solvent.

Acetylene: A hydrocarbon gas.

Acid: One of a class of substances which react with alkalis to give a salt and water.

Acrylonitrile: A monomer used to make Buna N.

Activated: Made more reactive.

Activators: Substances which improve the action of accelerators.

Addition Polymerization: Polymerization in which molecules simply add on to one another.

Ageing: The perishing or oxidation of rubber.

Alkali: A substance which reacts with an acid to produce a salt and water.

Aniline: An organic base. The first organic accelerator.

Antioxidants: Substances added to rubber to protect it from oxidation or ageing.

Atmospheric pressure: A pressure equal to the pressure of the atmosphere.

Atom: The smallest unit of a chemical element which can take part in a chemical reaction.

Autoclave: A vessel made to withstand high pressures.

Bakelite: A thermosetting plastic.

Bank: Excess rubber forming a mound above the nip on a mill.

Base: (See *Alkali.*)

Basic phase: The substance, in emulsion polymerization, in which the emulsion is made. Usually it is water.

Bead: The part of a tyre which holds it on to the rim.

Benzene (Benzole): A liquid hydrocarbon, which is an important constituent of some kinds of petrol.

Bias angle: The angle at which rubbered cord or fabric strips are cut, or the angle at which textile threads lie in the finished article.

Billet: A short thick rod.

Blanks: Pieces of compounded rubber cut approximately to size to fit into vulcanizing moulds.

Blow-down tank: A tank into which synthetic rubber latex is passed after leaving the reactor.

Bonds: Chemical links between atoms making up a compound atom or molecule.

Bowls: The rollers in a calender.

Brine: A strong solution of common salt.

B. rubber: German synthetic rubber made in World War I by polymerizing dimethyl butadiene with sodium in an atmosphere of carbon dioxide.

Buna N: A German special purpose synthetic rubber. A co-polymer of butadiene and acrylonitrile.

Buna rubber: The generic name given to a range of synthetic rubbers made in Germany between the wars, by polymerization of butadiene with sodium as the initiator. Later adopted as the name for all German synthetic rubbers containing butadiene.

Buna S: A co-polymer made from approximately 75 per cent butadiene and 25 per cent styrene.

Butadiene: A hydrocarbon, like isoprene. A monomer for S.B.R., nitrile rubber and polybutadiene.

Butt jointing: A process by which the ends of a tube are joined together to make the inner tube of a tyre.

Butylene: A hydrocarbon similar to butadiene.

Butyl rubber: A special purpose synthetic rubber made from isobutylene and a small amount of isoprene.

Calcium carbide: A chemical which reacts with water, giving off acetylene gas.

Calender: A machine for making thin sheets of rubber, coating fabric with rubber, or embossing rubber.

Camber: Lengthwise curvature of calender bowls to compensate for the distortion caused by rubber coming through the nip.

Carbon (*C*)*:* A chemical element present in coal, rubber, petroleum and many other materials.

Carbon black: A kind of refined soot.

Carbon dioxide: A gas which occurs in the atmosphere.

Carcase: (See *Casing*.)

Casing: The inner, textile reinforced part of a tyre cover.

Catalyst: A substance which enables a chemical reaction to proceed faster.

Catalytic cracking: Cracking effected by bringing hydrocarbons into contact with a catalyst.

Caustic soda: A strong alkali.

Centigrade: A scale of temperature. 0° Centigrade is the freezing point of water, and 100° Centigrade the boiling point.

Centrifuging: A method of concentrating latex.

Chafer strips: Narrow strips of rubberized canvas fitted on the bead region of a tyre.

Channel black: Carbon black made by burning natural gas and cooling the flame on a cold channel.

Chemical change: A change resulting in a substance with different chemical properties from the original.

Chemical property: The ability of a substance to change to something else with different chemical properties.

Chicle: A low-grade Central American rubber formerly used in making chewing gum.

2–Chlorobutadiene (chloroprene): The monomer from which Neoprene is made.

cis Configuration (of polyisoprene): All the CH_2 groups lie on one side of the rubber chain.

Clay: An example of a mineral silicate.

Coagulation: The clotting of latex.

Coal tar: A by-product of coal gas manufacture.

Coal tar naphtha: A solvent made from coal tar.

Cold flow: A property of silicone rubbers, whereby they slowly flow and take up the shape of a container.

Compound: A substance containing different kinds of chemical elements.

Compound (rubber): Raw rubber into which other substances have been mixed.

Compound atom: Another name for a molecule.

Compounding: Making a rubber compound.

Compression set: The amount by which rubber fails to come back to its original position after being released from compression.

Condensation polymerization: Polymerization in which a molecule is eliminated when two others join up.

Co-polymer: A polymer made by polymerizing two monomers together.

Cover: The part of a tyre which lies outside the inner tube.

Cracker: A special, robust mill used for breaking down rubber quickly. It is also used in the reclaiming process.

Cracking: The breaking down of large hydrocarbon molecules in petroleum oil to smaller ones.

Creaming: The partial concentration of latex, to give a cream with a higher rubber content.

Crêpe: A type of raw natural rubber.

Cross-links: Chemical links or bonds, joining molecular chains together, to effect vulcanization.

Crown: (See *Camber*.)

Crumb: Rubber just after it has been coagulated, or ground vulcanized rubber.

Curing: Another name for vulcanizing.

Dehydrogenation: A reaction involving the removal of hydrogen from a substance.

Die: A perforated plate at the end of the extruder through which rubber is forced.

Dimethyl butadiene: The monomer of the early German H and W rubbers.

Distillation: Heating a substance until it boils, then cooling and condensing the vapour.

Doctor blade: The knife-blade which helps to spread rubber solution onto cloth in the spreading machine.

Double bond: A reactive type of chemical bond which enables rubber to be vulcanized.

Dry distillation: Distillation of dry material.

Duprene: An early name for Neoprene.

Dynamic properties: The properties of rubber in motion.

Ebonite: Another name for hard rubber.

Elastic braid: Knicker elastic.

Elastic cord: Used in hat elastic or as springs.

Elasticity: The property of being able to stretch, then spring back on release.

Elastic web: Used in making braces and belts.

Elastic yarn: Thin rubber thread, covered with textile material.

Element (chemical): A substance containing atoms of one kind only.

Emulsion: A dispersion of very fine globules of one liquid in another.

Emulsion polymerization: The polymerization of the monomer in an emulsion—usually in water.

Ethyl benzene: A hydrocarbon intermediate used in the manufacture of styrene.

Ethylene: A hydrocarbon gas used in making the plastic known as polyethylene or 'Polythene'.

Ethylene dichloride: Used in the manufacture of certain Thiokol rubbers.

Extenders: Materials mixed into rubber to replace or extend it.

Extruder: A machine in which rubber is forced out through a hole or holes, forming a continuous piece of definite cross-sectional shape.

Fatty acid: An acid derived from fats or oils.

Feeding hopper: A kind of chute by which materials can be fed into a machine.

Ferrous metal: Iron and sometimes cobalt or nickel.

Filler strips: Strips of fabric reinforced rubber wrapped round the bead, in the construction of a tyre.

Fillers: (See *Inert fillers* and *Reinforcing fillers.*)

Flash: (See *Spew.*)

Flash tank: A tank in which butadiene or other volatile monomers can be removed from synthetic rubber latex by releasing the pressure.

Flux: An additive which enables a substance to melt at a lower temperature.

Former: A solid replica of an article, for dipping into latex to acquire a coating of rubber.

Formula (chemical): A chemical shorthand, giving the types and number of atoms in the molecule of a chemical compound.

Formula (compounding): A list of chemicals in a mix.

Free Radical Initiator: The type of initiator commonly used in emulsion polymerization.

French chalk: A white powder, sometimes used as a filler and also as a dusting powder to prevent unvulcanized rubber surfaces sticking together.

Frictioning: The coating of material, usually a textile, in a calender; where the rubber is pressed between the threads of the material.

Furnace black: A type of carbon black made by burning hydrocarbon gas or oils in an enclosed space, for example in a furnace.

Gas black: A carbon black made from natural petroleum gas.

Glycol: A chemical used in making polyurethane rubber.

Gutta percha: A hard, horn-like substance which is used as the outer covering of golf balls. It has the same chemical composition as natural rubber.

Hardness: The resistance a rubber shows to deformation.

Hard rubber: Vulcanized rubber containing about 30 per cent of sulphur combined with the rubber.

Heat build-up: The heat generated in a tyre due to poor resilience.

High styrene resins: Reinforcing fillers often used in shoe-soling rubber.

Homogeneous: Of uniform composition.

Horton spheres: Large cylindrical containers for holding butadiene or other gases under pressure.

H rubber: A German synthetic rubber made in World War I by the polymerization of dimethyl butadiene at room temperature.

Hydraulic: Refers to machines operated by the pressure of a liquid.

Hydrocarbon: A chemical compound made up of the elements hydrogen and carbon only.

Hydrochloric acid: A very strong acid, sometimes known as Spirits of Salts.

Hydrogen (*H*): The lightest chemical element known.

Hydrogenation: A reaction involving the combination of hydrogen with a substance.

Hysteresis: A phenomenon shown where the backwards and forwards stress-strain curves are not the same.

Inert: Will not react readily with other substances.

Inert fillers: Fillers which do not improve the physical properties of a rubber.

Inhibitor: Usually a substance which inhibits polymerization.

Initiator: A substance which enables a polymerization reaction to take place.

Insulator: In the electrical sense, a substance which will not pass electricity.

Internal mixer (*Banbury mixer*): An enclosed machine for masticating and mixing substances into rubber.

Ionic Initiators: A class of polymerization initiators. Recent developments in them have resulted in the production of synthetic natural rubber.

Isobutylene: A hydrocarbon monomer used in making butyl rubber.

Isocyanate groups: Groups of atoms which are particularly reactive and which enable molecular chains to be linked together in the making of polyurethane rubber.

Isomerism: The existence of *cis* and *trans* configurations in a molecule.

Isoprene: A hydrocarbon. The parent substance of natural rubber and also a monomer in some kinds of synthetic rubber.

Last: A shaped piece of metal for holding part of a shoe in the process of manufacture.

Latex: An emulsion of a rubber hydrocarbon in water.

Lead press: A kind of extruder, in which hot lead is applied to the outside of moulded hose during manufacture.

Light distillate: Lower boiling fractions of crude petroleum oil.

Light naphtha: (See *Light distillate.*)

Liner: Textile material onto which freshly calendered rubber is run, to prevent the rubber surfaces sticking together.

Mandrel: A long rod or tube, on which hose may be built or vulcanized.

Mastication: The reduction of rubber to a soft dough-like condition to enable substances to be mixed into it.

Mechanicals: A name describing a wide range of moulded goods and including soles and heels, bathing caps, hot water bottles, cycle pedals, door-stops and tiles.

Metal oxide: A compound containing a metallic element and oxygen.

Methyl chloride: A volatile liquid used as a diluent in the making of butyl rubber.

Mica: A mineral silicate.

Mill: A machine for masticating rubber, mixing substances into it, and making sheets of rubber.

Mix: A mixture of rubber with various substances.

Modifier: A substance which helps to produce chemical chains of desired length in polymerization.

Modulus: The weight, in pounds per square inch or other convenient units, which will stretch a piece of rubber to a standard elongation.

Molecule: A compound atom. The smallest unit of a chemical compound.

Monomer: A small molecule which makes the repeating units in a polymer chain.

Monovinyl acetylene: Used in the preparation of chloroprene, which is then polymerized to form Neoprene.

Mould (vulcanizing): Usually two pieces of stout metal which fit together and enclose a space the size and shape of the final object required.

Mustard gas: A poison gas used in World War I.

Natrium: The German name for sodium.

Neoprene: A special purpose rubber containing the element chlorine.

Nip: The distance between the rolls on a mill or calender.

Nitrile group: The carbon atom and its adjacent nitrogen atom in acrylonitrile.

Nitrile rubber: A special purpose rubber containing the element nitrogen.

Nitrogen (N): A gaseous element found in the air.

Organic bases: Materials which contain the element carbon and which react with acids.

Oxidation: Applied to rubber, means the ageing of rubber.

Oxidizing agents: Substances which help to oxidize rubber.

Oxygen (O): A gaseous element found in the air.

Ozone: A molecular form of oxygen, which is very reactive. It is found in small quantities in the atmosphere and generated near some electrical machines.

Particle accelerators: Electrical machines for producing some forms of penetrating radiation.

Pelota: A game, something like 'Fives', played by the Basque people of the western Pyrenees.

Perbunan: Nitrile rubber made from 25 per cent acrylonitrile and 75 per cent butadiene.

Perbunan Extra: Nitrile rubber made from 35 per cent acrylonitrile and 65 per cent butadiene.

Permanent set: The amount by which rubber fails to come back to its original position after being released from extension.

Permeability: The property of allowing gases or liquids to diffuse through a substance.

Petrol: Motor spirit. Some of the lowest boiling hydrocarbon fractions obtained by the distillation of crude petroleum oil.

Petroleum: A thick oil of mineral origin which is the source of many hydrocarbons.

Physical change: A change involving no loss of chemical identity.

Physical property: The ability to undergo a change without losing chemical identity.

Pine tar: A popular softener for rubber mixes, obtained by the distillation of pine wood.

Pitch: Generally the residue from the distillation of coal.

Plantation rubber: Natural rubber obtained from specially cultivated trees.

Plastic: Dough- or putty-like. Also used to describe a class of materials which at some stage in their manufacture are plastic like rubber, but which in the final state are neither plastic nor elastic.

Plasticizers: (See *Softeners.*)

Plutonium: A chemical element, the radioactive form of which is used in atomic bombs.

Ply: A layer of rubberized fabric.

Pneumatic tyre: A tyre which can be inflated with air.

Pocket construction: Two or more plies are assembled separately before being used to build the cover of a tyre.

Polyester chain: A part of the chemical chain in some polyurethane rubbers.

Polyether chain: A type of chemical chain present in some polyurethane rubbers.

Polyethylene: More commonly known as 'Polythene'. An important plastic.

Polyisoprene: The chemical name for the natural rubber chain.

Polymer: A molecule, often a chain-like molecule, made up of repeating units.

Polymerization: The process whereby a long chain of repeating units (polymer) is made by chemically linking a large number of small molecules together.

Polymerization formula: The list of substances used in the polymerization reaction.

Polystyrene: The polymer obtained when styrene is polymerized alone.

Polysulphide rubbers: Another name for Thiokol rubbers.

Polyurethanes: A range of rubbers usually containing the urethane group.
Potash: Potassium hydroxide. A strong alkali.
Processing: Treatment.
P.V.C.: Polyvinyl chloride. A plastic.

Radioactivity: Spontaneous radiation given out by certain elements.
Radioisotopes: Radioactive forms of certain elements.
Ram: A device for exerting pressure.
Raw rubber: Unvulcanized and uncompounded rubber.
Reactants: Substances which react together.
Reaction tower: A tower in which reaction takes place.
Reclaimed rubber: Scrap vulcanized rubber which has been treated so that it behaves like unvulcanized rubber in that it can be remilled and incorporated into a rubber mix.
Reinforcing fillers: Fillers which reinforce or improve some of the physical properties of rubber.
Resilience: The rubbery property of springing back after being released from tension.
Resins: Usually naturally occurring thermoplastic materials.
Retarders: Substances which retard or reduce premature vulcanization, but which do not interfere with the actual vulcanization process.
Rotor: The part of an internal mixer which corresponds to the rolls in a mill.
Rubber compound: (See *Compound (rubber)*.)

Salt: A substance produced by the action of an acid on an alkali.
Sand: Disintegrated rock, usually silicate material.
S.B.R.: Styrene-butadiene rubber. Also known as general purpose synthetic rubber.
Scorching: Premature vulcanization.
Self-vulcanizing solution: Normally a solution with a very fast accelerator, enabling vulcanization to be effected without the application of heat, when the solution has been dried down.
Short or end stop: A substance added to terminate the polymerization reaction.
Silica: Silicon oxide.
Silicates: Certain minerals containing the elements silicon and oxygen.
Silicon: A chemical element similar in some respects to the element carbon.
Silicone rubbers: Rubbers containing chains of alternate silicon and oxygen atoms.
Single ply building: Building a cover one ply at a time.
Smoked sheet: A type of raw natural rubber.
Soaker line: Tubes in a furnace in which cracking takes place.

Sodium: A reactive metal element, used as an initiator in some systems of polymerization.

Sodium tetrasulphide: Used in the manufacture of certain Thiokol rubbers.

Softeners: Substances which assist in mastication, or which help to soften the rubber for further processing.

'*Soft*' *vulcanized rubber:* Rubber vulcanized with a little sulphur to the tough and elastic state.

Solution: The result of dissolving a substance in a solvent.

Solvent: A liquid which will dissolve a substance.

Spew: Rubber squeezed out between the halves of a mould during curing.

Spreading machine: A machine for spreading thick rubber solutions on to cloth.

Steam cracking: Cracking effected by mixing hot hydrocarbon gases with steam.

Stearic acid: An organic fatty acid. An important activator of rubber accelerators.

Stockholm tar: (See *Pine tar.*)

Stress-strain curve or graph: A diagram obtained by plotting the weights required to give certain elongations, against the elongation.

Stripped latex: Synthetic rubber latex from which unreacted monomers have been removed.

Stripping column: A column in which styrene is removed from S.B.R. latex.

Styrene: A hydrocarbon. One of the monomers in S.B.R.

Sump: A collecting vessel.

Synthetic natural rubber: The name given to a man-made copy of natural rubber.

Synthetic substance: Strictly a substance which has been built up or synthesized, the substance synthesized having exactly the same structure as the natural one. In connection with rubber, it usually denotes a substance having rubbery properties.

Tack: The property of rubber sticking to itself.

Tailor-made polymers: Polymers made to have specific properties.

Tank farm: The area where monomers are stored.

Tensile properties: Properties connected with the strength of rubber.

Tensile strength: The weight a sample of rubber requires to break it, expressed in pounds per square inch of cross-sectional area or other convenient units.

Thermal black: Carbon black made by decomposing hydrocarbon gases by injecting them into very hot chambers.

Thermal cracking: Cracking produced by heat alone.

Thermal stability: Stability to heat.

Thermoplastic: Softens on heating.

Thermosetting: Sets hard on heating.

Thiokol: A special purpose rubber containing the element sulphur.

Titanium dioxide: A metal oxide, generally used as a white pigment.

Top coating or topping: Putting a coat of rubber onto one side of a material by means of a calender.

trans Configuration (of *polyisoprene*)*:* Pairs of CH_2 groups lie on alternate sides of the rubber chain.

Tread: The part of the tyre which comes into contact with the road.

Turbines: Machines which assist in turning the energy of steam, etc., into electricity.

Turpentine: An oily liquid obtained by distilling the resin of pine trees. Used at one time as a solvent for natural rubber.

Ultra accelerators: Very fast-acting accelerators.

Ultra violet light: Light of shorter wavelength than visible light.

Urethane group: A group of atoms appearing in polyurethane rubbers.

Volatile: Easily vaporized.

Vulcanization: Usually means heating rubber with sulphur, but it can be effected by other chemical agents. It really means cross-linking the rubber and producing a change from a weak plastic substance to a strong elastic one.

Vulcanite: Another name for hard rubber.

Vulkollan: One of the first of the polyurethane rubbers.

Warp: Lengthwise strands in woven fabric.

Water molecule: H_2O.

Weft: Cross strands in woven fabric.

Wild rubber: Natural rubber obtained from trees growing wild, i.e. not cultivated. It comes mostly from Brazil and West Africa.

W rubber: A German synthetic rubber made in World War I by the polymerization of dimethyl butadiene in heated drums.

Zinc oxide: A metal oxide filler which has some reinforcing properties, and is also sometimes used to impart a white colour to rubber. Its most effective use in rubber is, however, as an activator of accelerators in vulcanization.

Index

Name Index

www.ingramcontent.com/pod-product-compliance
Lightning Source LLC
LaVergne TN
LVHW011520200926
848484LV00001B/84
9780820600956